新形态计算机专业创新型人才培养系列教材

Vue.js 前端框架开发实战

崔妍妍　姚颖筠　杨　嘉　主编

·成都·

图书在版编目（CIP）数据

Vue.js前端框架开发实战 / 崔妍妍，姚颖筠，杨嘉主编. — 成都：成都电子科大出版社，2024.8

ISBN 978-7-5770-0952-0

Ⅰ.①V… Ⅱ.①崔… ②姚… ③杨… Ⅲ.①网页制作工具—程序设计 Ⅳ.①TP392.092.2

中国国家版本馆CIP数据核字（2024）第047658号

Vue.js前端框架开发实战

Vue.js QIANDUAN KUANGJIA KAIFA SHIZHAN

崔妍妍 姚颖筠 杨 嘉 主编

策划编辑	段 勇 魏 彬	
责任编辑	魏 彬	
责任校对	刘 凡	
责任印制	梁 硕	

出版发行 电子科技大学出版社
　　　　　成都市一环路东一段159号电子信息产业大厦九楼　邮编 610051
主　　页 www.uestcp.com.cn
服务电话 028-83203399
邮购电话 028-83201495

印　　刷 廊坊市伍福印刷有限公司
成品尺寸 210mm×285mm
印　　张 17.75
字　　数 575千字
版　　次 2024年8月第1版
印　　次 2024年8月第1次印刷
书　　号 ISBN 978-7-5770-0952-0
定　　价 69.00元

版权所有，侵权必究

编委会

总 顾 问 许绍兵

顾 问 沙 旭　徐 虹　蒋红建

主 编 崔妍妍　姚颖筠　杨 嘉

副 主 编（排名不分先后）
　　　　　宁玉丹　黄 欣　黄山松　保明庚
　　　　　曹 意　骆书祥　陈小虎

编 者（排名不分先后）
　　　　　夏伦梅　孙立民　宋家言　胡德俊
　　　　　范海涛　向 飞

编审委员会（排名不分先后）
　　　　　王 胜　吴凤霞　李宏海　俞南生
　　　　　吴元红　陈伟红　郭 尚　江丽萍
　　　　　王家贤　刘 雄　邓兴华　王 志
　　　　　徐 磊　钱门富　陈德银　赵 华
　　　　　汪建军　陶方龙　尹 峰　杜长田
　　　　　费 群　芮贵锋　赵亚斌

专业委员（排名不分先后）
　　　　　盛文兵　范树明　范晓燕　王明鑫
　　　　　王 强　邹时桥　王丽萍　吴 锐
　　　　　王天德　茂文涛　黄超男　贾 进
　　　　　曹永军　华永红　何 健　王文杰
　　　　　张冠儒　侯海涛　赵世浩　张婷婷
　　　　　韩士权　罗靖宁　程全锋

前 言

党的二十大明确提出全面建设社会主义现代化国家的目标任务。我们意气风发迈向新征程。在党的二十大精神的指引下，我们将在网站前端开发领域中开启新的学习内容。

Vue.js 是当今最受欢迎的 JavaScript 框架之一，在现代 Web 开发中具有广泛的市场应用。本书共 9 个项目章节，详细讲解了 Vue.js 的使用方法。本书采用项目实战教学，案例丰富，有近百个案例，并且每一章都有综合项目案例。对于大项目案例采用分步骤讲解的方式，能有效降低读者学习的难度。根据职业教育"重在实践"的教学特点，以"知识要点＋项目实战＋微课教学"为主线，注重专业技能、动手能力的培养。

本书从 Vue.js 起步入手，结合大量的案例分析，对 Vue.js 知识点进行了细致的取舍和编排，深入讲解了 Vue.js 实例结构、Vue.js 指令、Vue.js 过渡与动画、Vue.js 请求网络数据、Vue.js 组件、Vue.js 路由。每个章节由若干个实战项目组成，并录制微课供读者自学。

本书融通俗性、实用性和技巧性于一身，适用于网站前端开发设计师、小程序开发设计师，以及相关专业的高校学生阅读。读者可以根据自己的需要选择不同的章节进行学习，并从中汲取灵感和思路。在学习本书的过程中，读者可以结合实际案例进行思考和实践，以达到最佳的学习效果。本书中的案例都有配套的源代码和相关素材，方便读者进行学习。为了便于书籍编排及展现，本书案例中的 JavaScript 代码采用内嵌方式，请读者知悉。

在本书的编写过程中，编委会中的各位领导和专家们给予了大力支持和指导，在此一并表示感谢！由于编者水平有限，书中难免有不妥之处，恳请广大读者谅解并提出宝贵的意见。

Vue.js 前端框架开发实战 – 课件二维码

目　　录

第 1 章　Vue.js 的起步 ·· 1
1.1　Vue.js 的简介及开发优势 ··· 2
1.2　MVVM 模式 ·· 3
1.3　安装 Vue.js ··· 4
1.4　创建第一个 Vue 实例 ··· 5
1.5　综合实训：在页面渲染三个变量 ··· 8

第 2 章　Vue 实例结构 ··· 11
2.1　Vue 实例结构的组成 ··· 12
2.2　数据的绑定 ·· 16
2.3　Vue 生命周期 ··· 23
2.4　综合实训Ⅰ：制作计数器效果 ··· 25
2.5　综合实训Ⅱ：制作逐字显示效果 ··· 27

第 3 章　Vue 内置指令 ··· 30
3.1　v-text、v-cloak、v-html 指令 ··· 31
3.2　v-on 指令 ·· 34
3.3　v-if、v-else、v-show 指令 ··· 39
3.4　v-for 指令 ··· 43
3.5　v-bind 指令 ··· 47
3.6　v-model 指令 ·· 54

第 4 章　Vue 内置指令综合演练 ··· 62
4.1　工作计划表 ·· 63
4.2　图书管理系统 ··· 71
4.3　焦点图效果 ·· 84

第 5 章　Vue 过渡与动画 ·· 93
5.1　Vue 过渡 ··· 94
5.2　Vue 动画 ··· 98

5.3 群组对象添加过渡或动画 ··········· 102

第 6 章 Axios 的使用 ··········· 110
6.1 使用 Axios 发起网络请求 ··········· 111
6.2 通过经纬度数获取地理信息 ··········· 118
6.3 获取随机背景图 ··········· 124
6.4 获取城市天气 ··········· 128

第 7 章 Vue.js 组件 ··········· 140
7.1 组件的基本结构 ··········· 141
7.2 组件的切换 ··········· 148
7.3 组件的通信 ··········· 154
7.4 插槽 ··········· 165
7.5 综合实训：使用组件实现购物车功能 ··········· 170

第 8 章 Vue.js 路由 ··········· 185
8.1 路由基础 ··········· 186
8.2 路由传参数 ··········· 198
8.3 综合实训：茶叶网站的制作 ··········· 207

第 9 章 Vue CLI ··········· 228
9.1 Vue CLI 简介及安装 ··········· 229
9.2 Vue CLI 初始化项目介绍 ··········· 235
9.3 使用 Vue CLI 完成品茶轩项目 ··········· 239

参考文献 ··········· 273

参考答案 ··········· 274

第 1 章

Vue.js 的起步

导言

中国 IT 行业已经成为全球最大的互联网市场之一，近年来取得了显著的发展成就。电子商务的繁荣为我们带来巨大的市场效益。移动互联网的普及促进了移动支付、在线教育等领域的快速发展。人工智能的兴起对我们的工作方式和产业结构产生了深远的影响。IT 行业创新创业的蓬勃发展为我国经济注入了新的活力。"长风破浪会有时，直挂云帆济沧海"。在本章中我们将步入 Vue.js 的学习。希望同学们能在 IT 发展的浪潮中实现自身的理想和价值。

学习内容

本章是 Vue.js 的起步章节，一共有 5 节。1.1 节主要讲解 Vue.js 的开发优势。1.2 节讲解 MVVM 模式。1.3 节讲解 Vue.js 的三种安装方式。1.4 节分别使用 Vue 2.0 和 Vue 3.0 两个不同的版本来创建 Vue 实例，通过实例来对比 Vue 2 和 Vue 3 在使用上的区别。1.5 节通过一个综合案例来进行巩固和练习。

学习目标

1. 了解 Vue.js 的特点以及 Vue.js 在前端开发中的优势。
2. 掌握 MVVM 模式的组成。
3. 掌握 Vue.js 的安装方法。
4. 掌握 Vue 2 和 Vue 3 实例化 Vue 对象、数据（data）和方法（methods）的写法。

学习重点

1. 掌握 MVVM 模式包含的三个部分。
2. 掌握 Vue.js 的本地安装方法。
3. 掌握 Vue 3 实例挂载、data、methods 的写法。

1.1　Vue.js 的简介及开发优势

1.1.1　Vue.js 简介

Vue.js 于 2014 年发布，它的创建者是 Evan You，中文姓名为尤雨溪。Vue.js 是基于标准 HTML、CSS 和 JavaScript 构建的，它是一套用于构建用户界面的渐进式框架。渐进式的概念是指 Vue.js 是分层设计的，每层可选，不同层可以灵活接入其他方案架构模式或者第三方库。当开发者对 Vue.js 的掌握度不高时可以只使用 Vue.js 做页面渲染、表单处理提交等基础操作，不需要引入其他复杂的功能。当项目规模逐渐变大，就可以逐步引入前端路由、状态集中管理等相对复杂的功能。Vue.js 的官网地址为 https://cn.vuejs.org/，如图 1-1 所示。

图 1-1　Vue.js 官网

1.1.2　Vue.js 开发优势

Vue.js 是当今最受欢迎的 JavaScript 框架之一，在现代 Web 开发中具有广泛的市场应用。Vue.js 的开发优势有以下几点：

（1）轻量级框架、简单易学、运行效率高，适合移动 /PC 端开发；

（2）遵循 MVVM 模式，视图、数据、结构分离；

（3）可以对数据进行双向绑定，开发者不需要通过操作 DOM 对象，数据发生变化时页面就发生变化；

（4）Vue.js 可以进行组件化开发，方便代码复用，同时使代码编写量大大减少；

（5）Vue.js 使用路由可以实现单页面应用，使页面局部刷新，响应速度快。而传统的页面通过超链接实现页面的跳转，每次跳转页面都要请求所有数据和 DOM。

1.2 MVVM 模式

1.2.1 MVVM 模式介绍

MVVM 是 Model（模型）-View（视图）-View Model（视图模型）的简写，即由 M-V-VM 三部分组成。如图 1-2 所示。

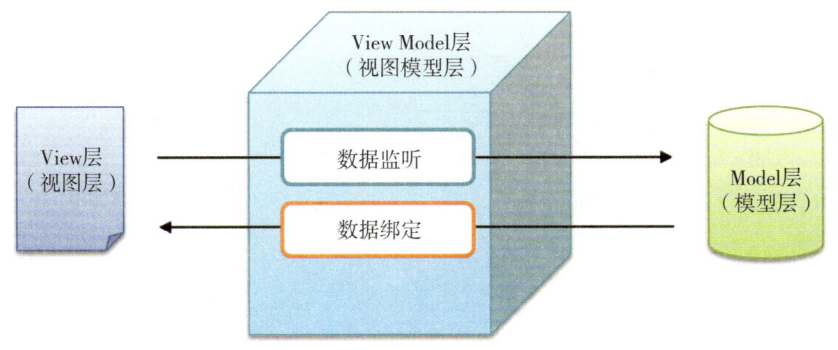

图 1-2 MVVM 模式介绍

Model 层是模型层。模型层负责数据的获取、存储、处理和验证等。模型层中的数据可以是固定的数据，也可以是来自服务器或者从网络上请求下来的数据。

View Model 层是视图模型层，它是 View 和 Model 沟通的桥梁，是实现数据双向绑定的关键。一方面它实现了将 Model 的改变实时反映到 View 中，另一方面它实现了 DOM 监听，当 DOM 发生一些事件，如点击、滚动等，可以监听到这些事件，并在需要的情况下改变对应的数据。

View 层是视图层。在前端开发中对应的是 DOM 层，主要作用是通过 HTML 结构给用户展示各种信息。

MVVM 开发模式采用了数据绑定的方式，减少了与 UI 交互的代码量。MVVM 开发模式将用户界面的开发和业务逻辑的开发分离开，因此可以更轻松地进行各部分的测试和开发，同时由于不同部分彼此独立，也更加利于开发人员之间分配任务和彼此协作。

1.2.2 MVVM 实例

下面通过具体的实例来展示 MVVM 模式。MVVM 实例如图 1-3 所示，在浏览器中点击 "+" 号按钮可以让示数增大，点击 "-" 号按钮可以让示数减少。该实例中购买数量是通过渲染数据得到的，在控制台中查看结构可以看到，当数据增大或者减少时，不影响网页的 DOM 结构，只是购买数量的值进行了更新，如图 1-4 所示。

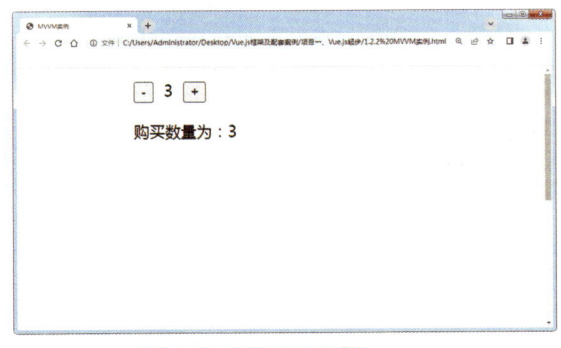

图 1-3 MVVM 实例

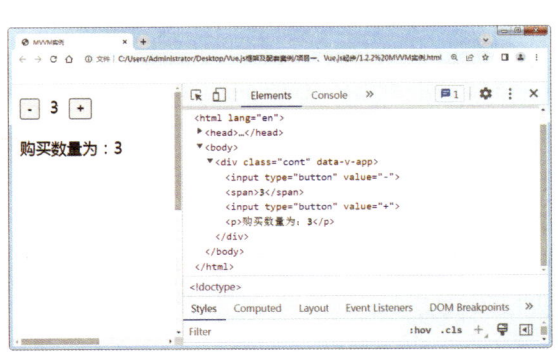

图 1-4 HTML 结构

模型层（Model 层）对应的代码如下：

```
data(){
    return {
        num:0
    }
},
```

视图层（View 层）对应的代码如下：

```
<div class="cont">
    <input type="button" value="-" @click="reduce()">
    <span> {{num}} </span>
    <input type="button" value="+" @click="add()">
    <p>购买数量为：{{num}}</p>

</div>
```

视图模型层（View Model 层）是 Vue.js 实例，对应的代码如下：

```
const app={
    data(){
        return {
            num:0
        }
    },
    methods:{
        add(){
            this.num++
        },
        reduce(){
            if(this.num>0){
                this.num--
            }
        },
    }
}
Vue.createApp(app).mount('.cont')
```

1.3 安装 Vue.js

Vue.js 的安装方法有多种，下面列举三种安装方法：使用独立版本安装，使用 CDN 方法安装，使用 NPM 方法安装。

1.3.1 使用独立版本安装

从互联网上下载 Vue.js 文件到本地,然后使用 <script> 标签进行引用。这种方法将 Vue.js 文件放存在本地目录下,使用时不需要连接互联网,推荐初学者使用该方法。

安装代码如下:

```
<script src="./js/vue3.js"></script>
```

1.3.2 使用 CDN 方法

在 <script> 标签中链接网络地址。使用该方法时需要连接互联网,但是不需要将 Vue.js 文件下载到本地。Vue 官网提供的链接地址如下:

```
<script src="https://unpkg.com/vue@3/dist/vue.global.js"></script>
```

1.3.3 使用 NPM 方法安装

使用 NPM 方法安装 Vue.js 时要求本地已安装 16.0 或更高版本的 Node.js,然后在命令行窗口中运行 npm init vue@latest,即安装 Vue.js 的最新版本。这一指令将会安装并执行 create-vue,它是 Vue 官方的项目脚手架工具。

1.4 创建第一个 Vue 实例

Vue 目前最新的是 Vue 3 版本。Vue 官网中声明:Vue 2 将于 2023 年 12 月 31 日停止维护,但是 Vue 2 版本广泛存在于已有的项目中,所以 Vue 2 版本的语法也要熟悉。下面的案例中将演示 Vue 2 和 Vue 3 两种版本在创建 Vue 实例上的区别。

说明:本书中的案例没有特殊说明的情况下使用的都是 Vue 3 版本。

1.4.1 使用 Vue 2 创建 Vue 实例

使用 Vue 2 创建 Vue 实例

```html
<!DOCTYPE html>
<html lang="en">
<head>
    <meta charset="UTF-8">
    <meta http-equiv="X-UA-Compatible" content="IE=edge">
    <meta name="viewport" content="width=device-width, initial-scale=1.0">
    <title>使用 Vue2 创建 Vue 实例</title>
    <script src="./js/vue2.js"></script>

    <script>
        window.onload=function(){
            new Vue({
```

```
            el:".container",
            data:{
                msg:"hello Vue!"
            },
            methods:{
                introduce(){
                    alert('易学易用,性能出色,适用场景丰富的 Web 前端框架!')
                }
            }
        })
    }
    </script>

</head>
<body>
    <div class="container">
        <h1 @click="introduce()">{{msg}}</h1>
    </div>
</body>
</html>
```

".container" 对应的是 HTML 结构中的 <div class="container">

其中,new Vue() 的作用是创建一个 Vue 实例,每一个 new Vue() 都是一个 Vue 构造函数实例。

创建实例时,需要传入选项对象。选项对象包括 el(挂载元素)、data(数据)、methods(方法)等。

el:". container" 设置的是该 Vue 实例针对的 HTML 根容器的元素,如果在 HTML 结构中使用的是 <div id="container"></div>,则应当使用 el:"#container"。

data 设置的该 Vue 实例中的变量。

methods 设置的是该 Vue 实例中的方法,在 h1 标签中使用 @click 即点击的时候来调用该方法。

实例效果为:在 h1 里面渲染出 data 中 msg 对应的数据,点击 h1 出现弹窗,效果如图 1-5 所示。

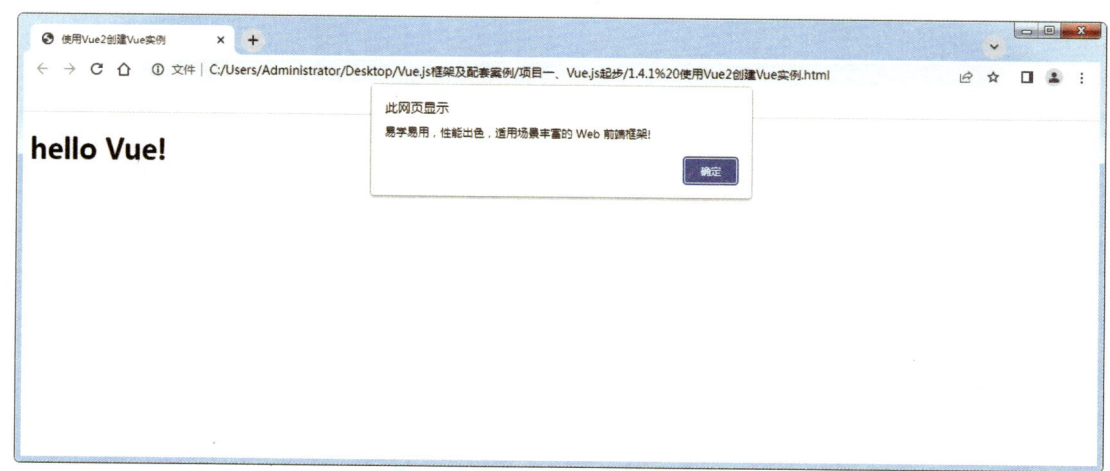

图 1-5　Vue 2.0 实例

1.4.2 使用 Vue 3 创建 Vue 实例

```html
<!DOCTYPE html>
<html lang="en">
<head>
    <meta charset="UTF-8">
    <meta http-equiv="X-UA-Compatible" content="IE=edge">
    <meta name="viewport" content="width=device-width, initial-scale=1.0">
    <title>Document</title>
    <script src="./js/vue3.js"></script>
    <script>
        window.onload=function(){
            const   app={           // 定义一个 Vue 实例 "app"
                data(){
                    return {        // Vue 3.0 中的变量必须使用返回值的形式来定义
                        msg:"hello vue!"
                    }
                },
                methods:{
                    introduce(){
                        alert('易学易用,性能出色,适用场景丰富的 Web 前端框架!')
                    }
                }
            }
            Vue.createApp(app).mount('#container')    // 将实例挂载到 #container
        }
    </script>
</head>
<body>
    <div id="container">
        <h1 @click="introduce()">{{msg}}</h1>
    </div>
</body>
</html>
```

使用 Vue 3.0 创建实例的方法是:首先定义一个 Vue 实例 "app",然后使用 createApp 来创建应用,调用 mount 方法传入选择器将实例挂载在 #container 这个 dom 节点上。

在 Vue 3 中,data 的数据必须以函数返回值的形式出现,即必须写在 return 中,如果依然写成 Vue 2 中,data 的形式则会出现语法错误。在 Vue 3 中 methods 的写法与 Vue 2 中是相同的。

实例效果与 Vue 2 中的实例效果相同。在 h1 里面渲染出 data 中 msg 对应的数据，点击 h1 出现弹窗，如图 1-6 所示。

图 1-6　Vue 3.0 实例

1.5　综合实训：在页面渲染三个变量

1.5.1　项目描述

该案例使用 Vue 3 创建一个 Vue 实例，在 VS Code 或者其他编译器中编写代码，在页面中渲染出三个变量。在 data 中定义三个变量如下：

`title:"Vue.js"`，

`subtitle:"渐进式 JavaScript 框架"`，

`des:"易学易用，性能出色，适用场景丰富的 Web 前端框架。"`

分别使用 h1、h3 和 p 标签将三个变量在页面中渲染出来。完成后的效果如图 1-7 所示。

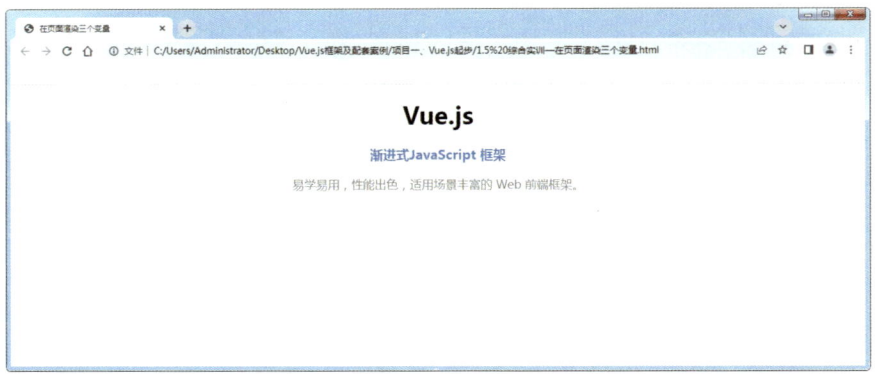

图 1-7　在页面渲染三个变量

1.5.2　项目分析

首先引用 Vue.js，然后创建 Vue 3.0 实例，在 data 中定义变量，在 HTML 结构中将变量渲染出来。

1.5.3　项目实施

在样式中对 h1、h3 和 p 标签进行设置，在 data 中定义三个变量，然后在 HTML 结构中分别使用

h1、h3 和 p 标签将三个变量渲染出来，具体代码如下：

```html
<!DOCTYPE html>
<html lang="en">
<head>
    <meta charset="UTF-8">
    <meta http-equiv="X-UA-Compatible" content="IE=edge">
    <meta name="viewport" content="width=device-width, initial-scale=1.0">
    <title>在页面渲染三个变量</title>
    <style>                                              /* 设置样式 */
        body{ font-family:"微软雅黑";}
        h1{ text-align: center;font-size: 36px;}
        h3{text-align: center;font-size: 20px; color: cornflowerblue;}
        p{ text-align: center; font-size: 18px; color: #999;}
    </style>
    <script src="./js/vue3.js"></script>
    <script>
        window.onload=function(){
            const app={
                data(){                                  /* 定义变量 */
                    return {
                        title:'Vue.js',
                        subtitle:'渐进式 JavaScript 框架',
                        des:'易学易用，性能出色，适用场景丰富的 Web 前端框架。'
                    }
                },
            }
            Vue.createApp(app).mount('#container')
        }
    </script>
</head>
<body>
    <div id="container">
        <h1>{{title}}</h1>                               <!-- 将变量渲染到页面 -->
        <h3>{{subtitle}}</h3>
        <p>{{des}}</p>
    </div>
</body>
</html>
```

课后练习题

1. MVVM 是_____的简写，_____层是模型层，_____层是视图模型层，_____层是视图层。

2. 从互联网上下载 Vue.js 文件到本地，使用_____标签进行引用。

3. 使用 Vue 2.0 创建一个 Vue 实例时需要传入选项对象。选项对象包括_____（挂载元素）、_____（数据）、_____（方法）等。

4. 使用 Vue 3.0 创建实例的方法是：首先定义一个 Vue 实例，然后使用_____来创建应用，调用_____方法传入选择器将实例挂载在 #container 这个 dom 节点上。

5. 在 Vue 3 中，data 的数据必须以_____的形式出现，即必须写在 return 内部。

第 2 章

Vue 实例结构

导言

平凡的生活，平凡的故事，却有最深的感动。在我们的国家，每天都有无数的平凡人，步履不怠，奋斗不止，上演着许多平凡又不平凡的故事，绘就了国家的繁荣与希望。本章我们将学习 Vue 实例的创建。Vue 实例由多个部分组成，每一部分都有各自的功能。我们每个人都是祖国建设者中的一分子，让我们用努力和汗水筑梦祖国美好的未来。

学习内容

本章主要讲解 Vue 的实例结构，一共有 5 节。2.1 节是 Vue 实例结构的概述。2.2 节讲解数据的绑定、计算属性 computed 和数据监听 watch 的使用方法。2.3 节讲解 Vue 生命周期的钩子函数。2.4 和 2.5 节是两个综合案例，分别是制作计数器效果和逐字显示效果。

学习目标

1. 了解 Vue 实例的组成部分。
2. 掌握 data 中数据的类型。
3. 掌握插值表达式的使用。
4. 掌握 computed 的使用方法。
5. 掌握 Vue 生命周期的钩子函数。

学习重点

1. 掌握 data 的数据类型。
2. 掌握插值表达式的使用方法。
3. 掌握 computed 的使用方法。
4. 掌握 created 的使用方法。

2.1　Vue 实例结构的组成

2.1.1　Vue 实例的基本组成

Vue 实例的基本组成包括数据（data）、方法（methods）和挂载的 DOM 结构。下面的 Vue 实例中，在 data 中定义了变量 msg，在 methods 中定义了 hello() 函数，具体代码如下：

```html
<!DOCTYPE html>
<html lang="en">
<head>
    <meta charset="UTF-8">
    <meta http-equiv="X-UA-Compatible" content="IE=edge">
    <meta name="viewport" content="width=device-width, initial-scale=1.0">
    <title>Vue 基本结构 </title>
    <script src="./js/vue3.js"></script>
    <script>
      window.onload=function(){
        const app={
          data(){              // 在 data 中以返回值的方式来存放变量
            return {
              msg:" hello vue!",
            }
          },

          methods:{            // 在 methods 中存放方法
            hello:function(){
              alert('易学易用，性能出色，适用场景丰富的 Web 前端框架。')
            },
          }
        }
        Vue.createApp(app).mount('#container')   // 创建 vue 实例挂载到 <div id="container"></div> 上
      }
    </script>
</head>
<body>
    <div id="container">
        <h2 @click="hello()">{{msg}}</h2>
    </div>
```

```
</body>
</html>
```

在 data 中定义变量的格式为：

```
msg:"hello vue!",
```

其中，msg 为变量名，"hello vue!" 为变量 msg 所对应的值。一个变量定义完成后，使用英文逗号（,）间隔下一个变量。

在 methods 中定义方法的格式为：

```
hello:function(){
            alert('易学易用，性能出色，适用场景丰富的 Web 前端框架。')
        },
```

其中，hello 为函数名称，hello:function( ){} 的格式也可以写成 hello( ){}。一个函数定义完成后，使用逗号（,）间隔下一个函数。

加载页面后，变量 msg 会渲染到 h2 标签中，点击 h2 标签可以弹出一个警示框，如图 2-1 所示。

图 2-1　Vue 实例简单结构

2.1.2　Vue 实例的复杂组成

Vue 实例中除了 data 和 methods，有时候还会有其他的结构，下面案例中使用了计算属性和生命周期的 create 函数，后面会有这些结构的详细讲解，这里只需要关注其结构的格式即可。

```
<!DOCTYPE html>
<html lang="en">
<head>
    <meta charset="UTF-8">
    <meta http-equiv="X-UA-Compatible" content="IE=edge">
    <meta name="viewport" content="width=device-width, initial-scale=1.0">
    <title>Vue 实例的复杂组成</title>
    <script src="./js/vue3.js"></script>
```

```html
<script>
  window.onload=function(){
    const app={
      data(){
        return {
          msg1:'hello vue!',
          msg2:'I like vue.js!',
          xing:'',
          ming:'',
        }
      },
      methods:{
        hello1(){
          alert('渐进式JavaScript框架')
        },
        paomadeng(){
          setInterval(()=>{
            const start=this.msg2.slice(0,1)
            const end=this.msg2.slice(1,)
            this.msg2=end+start
          },500)



        }

      },
      computed:{     ← computed 是计算属性，与 methods 和 data 同级别
        name(){
          return this.xing+this.ming;
        }
      },
      created(){     ← created 是生命周期函数，与 methods 和 data 同级别
        this.paomadeng()
      }
    }
    Vue.createApp(app).mount('#container')
  }
</script>
</head>
```

```
<body>
    <div id="container">
        <div class="name">
            <p>请输入您的姓：<input type="text" v-model="xing" ></p>
            <p>请输入您的名：<input type="text" v-model="ming" ></p>
        </div>
        <h3>你好,{{name}}</h3>
        <p>{{msg2}}</p>
        <h2 v-on:click="hello1()">{{msg1}}</h2>
    </div>
</body>
</html>
```

案例执行效果如图 2-2 所示。

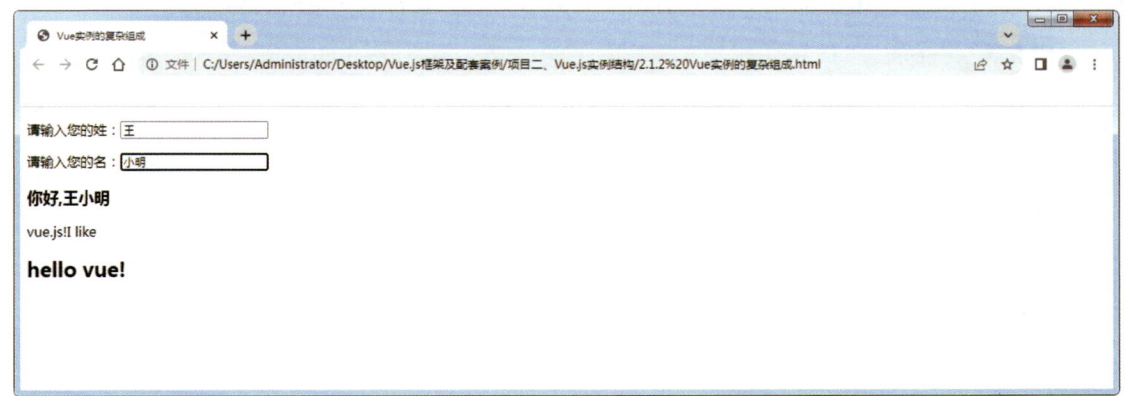

图 2-2　Vue 实例复杂结构

2.1.3　Vue 数据类型

Vue 数据类型可以分为数字型、字符型、布尔型、数组类型、对象类型等。例如：

```
data(){
    return {
        msg:"hello vue.js",
        num:10,
        flag:true,
        fruit:["香蕉","苹果","橘子"],
        student:{
            name:"小明",
            age:17,
            score:{
                Chinese:95,
                Math:98,
```

```
                English:96,
            }
        }
    }
}

msg:"hello vue.js",
```

变量 msg 是字符串类型变量,字符串类型的变量需要在值的两边使用引号,单引号(')或者双引号(")都可以。

```
num:10,
```

num 是数字类型变量,可以用于加减乘除等数学计算。

```
flag:true,
```

flag 是布尔类型变量,布尔类型的值只有两个:真(true)或者假(false)。

```
fruit:["香蕉","苹果","橘子"]
```

fruit 是数组类型变量,fruit[0] 对应的值是" 香蕉 ",fruit[1] 对应的值是" 苹果 ",fruit[2] 对应的值是" 橘子 "。

```
student:{
        name:"小明",
        age:17,
        score:{
                Chinese:95,
                Math:98,
                English:96,
                
            }
        }
```

student 是对象型变量。在对象型变量中使用键值对的形式来记录变量,使用点(.)运算来获取对象内部的值。student.name 对应的值是" 小明 ",student.score.Math 对应的值是 98。

2.2 数据的绑定

插值表达式

2.2.1 插值表达式

数据的绑定分为单向绑定和双向绑定,双向绑定使用 Vue 指令 v-model 来实现,在后面章节中会详细讲解,这里主要讲解单向绑定。数据的单向绑定使用插值表达式,符号为 {{ }}。例如下面代码。

```html
<!DOCTYPE html>
<html lang="en">
<head>
    <meta charset="UTF-8">
    <meta http-equiv="X-UA-Compatible" content="IE=edge">
    <meta name="viewport" content="width=device-width, initial-scale=1.0">
    <title>插值表达式</title>
    <script src="./js/vue3.js"></script>
    <script>
        window.onload=function(){
            const app={
                data(){
                    return {
                        msg:"hello vue.js",//字符串类型变量
                        num:10,//数字类型变量
                        flag:true,//布尔类型变量
                        fruit:["香蕉","苹果","橘子"],//数组类型变量
                        student:{//对象类型变量
                            name:"小明",
                            age:17,
                            score:{
                                Chinese:95,
                                Math:98,
                                English:96,
                            }
                        }
                    },
                }
            }
            Vue.createApp(app).mount('#container')
        }
    </script>
</head>
<body>
    <div id="container">
        <h2>{{msg}}</h2>
        <p>num的值是:{{num}}</p>
        <p>flag的值是:{{flag}}</p>
```

```
        <p>水果的种类有:{{fruit}},最喜欢的水果是:{{fruit[2]}}</p>
        <p>{{student.name}}的年龄是{{student.age}},他的数学分数是{{student.score.Math}}</p>
    </div>
</body>
</html>
```

代码执行后的效果如图 2-3 所示。

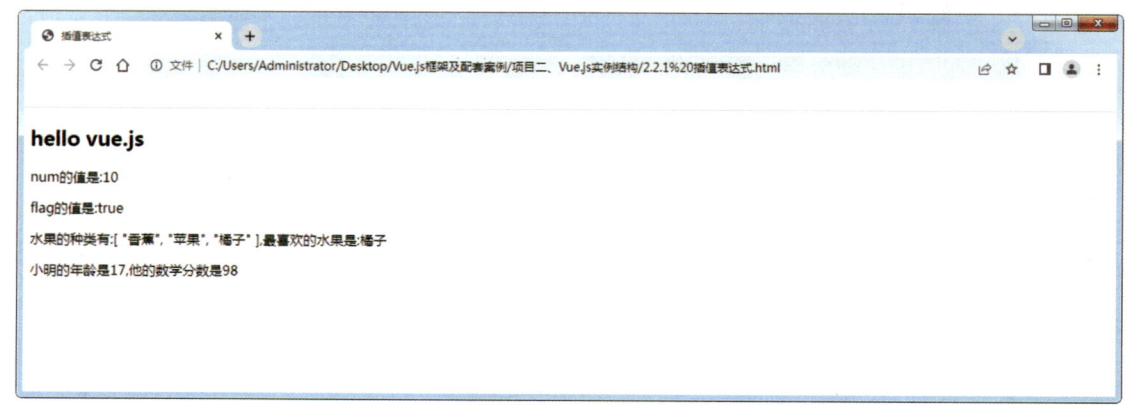

图 2-3　插值表达式

插值表达式的使用方法是将变量放在 {{ }} 之间。差值表达式可以直接和标签以及文字进行混排，中间不需要连接符号。

2.2.2　this 的使用

变量直接放在插值表达式中就可以渲染出来，但是在函数中如果使用 data 中的变量则需要在变量前面添加 this。这里的 this 指代的是该 Vue 实例对象。在 Vue 实例中，一个函数如果要引用 methods 中定义的函数，也需要在这个函数的前面添加 this。在下面案例中，data 中有一个变量 name，当点击 h2 标签时，可以获取这个变量的值。

```
<!DOCTYPE html>
<html lang="en">
<head>
    <meta charset="UTF-8">
    <meta http-equiv="X-UA-Compatible" content="IE=edge">
    <meta name="viewport" content="width=device-width, initial-scale=1.0">
    <title>this 的使用</title>
    <script src="./js/vue3.js"></script>
    <script>
      window.onload=function(){
        const app={
          data(){
            return {
              name:'小明',
```

```
                    }
                },
                methods:{
                    hello:function(){
                        alert(this.name+'你好!')
                    },
                }
                Vue.createApp(app).mount('#container')
            }
        </script>
    </head>
    <body>
        <div id="container">
            <h2 @click="hello()">{{name}}</h2>
        </div>
    </body>
</html>
```

变量前添加 this

代码执行后的效果如图 2-4 所示。

图 2-4　this 的使用

2.2.3　计算属性 computed

计算属性使用 computed 来进行定义，在 HTML 结构中也是使用 {{ }} 来显示计算的结果。但是它与一般的变量不同，它的本质是一个方法，在里面写一些计算逻辑的属性，然后使用 return 返回一个结果。computed 具有缓存特性，在 computed 中使用 data 中的某些变量，只有当这些变量发生变化时，计算属性嗅探到这种变化，才会自动执行 computed，这样就能够避免每次获取值时都要重新计算。

下面是使用 computed 来实现字符反转的案例。

```html
<!DOCTYPE html>
<html lang="en">
<head>
    <meta charset="UTF-8">
    <meta http-equiv="X-UA-Compatible" content="IE=edge">
    <meta name="viewport" content="width=device-width, initial-scale=1.0">
    <title>Document</title>
    <script src="./js/vue3.js"></script>
    <script>
        window.onload=function(){
            const app={
                data(){
                    return {
                        str:"ABCD",
                    }
                },
                methods:{
                    change(){
                        this.str="1234"
                    }
                },
                computed:{
                    reserveMsg(){
                        return this.str.split("").reverse().join("")
                    }
                }
            }
            Vue.createApp(app).mount('#container')
        }
    </script>
</head>
<body>
    <div id="container">
        <p>原字符串：{{str}}</p>
        <p>反转以后的字符串：{{reserveMsg}}</p>
        <input type="button" value="换一换" @click="change()">
    </div>
</body>
</html>
```

代码执行后的效果如图 2-5 所示,点击"换一换"按钮后的效果如图 2-6 所示。

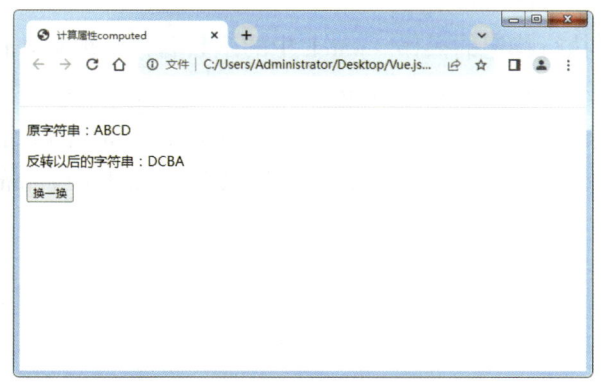

图 2-5　初始状态

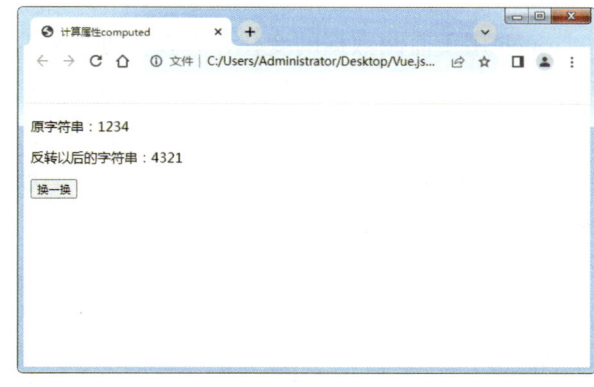

图 2-6　点击"换一换"

在上述代码中,split( ) 的作用是把一个字符串切割成字符串数组,reverse( ) 的作用是颠倒数组中元素的顺序,join( ) 的作用是把数组中的所有元素连接为一个新的字符串。

computed 中定义了 reserveMsg( ),在 reserveMsg( ) 函数中通过一系列方法将字符串进行反转。在 HTML 结构中使用 {{reserveMsg}} 来渲染数据。当点击"换一换"按钮,将 str 变量的值变成"1234"时,computed 可以监听到 str 值的改变,会立即将新的 str 变量进行反转。

在 computed 中不能对 data 中的同名变量进行赋值操作。如果在 data 中定义变量 reserveMsg,执行 computed 会出现报错。如图 2-7 所示。

```
data(){
    return {
        str:"ABCD",
        reserveMsg:",// 在 data 中定义同名称的变量

    }
}
```

图 2-7　控制台报错

2.2.4　watch 的使用

watch 用于监测数据的变化，并在数据发生改变时执行特定的操作。watch 中的函数名称必须和 data 中的某个数据名称一致，当这个值发生变化时就会触发某些业务功能。watch 不支持缓存，watch 中的函数也不需要调用。watch 中的函数会有两个参数：一个是新的值，一个是原来的值。下面是 watch 的演示案例，案例中 count 的初始值为 0，当点击"增加 1"按钮时，count 的值发生改变，触发 watch 中 count() 函数。具体代码如下：

```html
<!DOCTYPE html>
<html lang="en">
<head>
    <meta charset="UTF-8">
    <meta http-equiv="X-UA-Compatible" content="IE=edge">
    <meta name="viewport" content="width=device-width, initial-scale=1.0">
    <title>watch</title>
    <script src="./js/vue3.js"></script>
    <script>
        window.onload=function(){
            const app={
                data(){
                    return {
                        count:0,
                    }
                },
                methods:{
                    add(){
                      this.count++
                    }
                },
                watch:{
                    count(newVal,oldVal){
                        console.log("新的数值是"+newVal+",旧的数值是"+oldVal)
                    }
                }
            }
            Vue.createApp(app).mount('#container')
        }
    </script>
</head>
<body>
    <div id="container">
```

```
            <p>count 的值是 :{{count}}</p>
            <input type="button" value="增加 1" @click="add()">
        </div>

    </body>
</html>
```

页面加载后的效果如图 2-8 所示，每当点击"增加 1"按钮时，count 的值会发生改变，既而触发一次 count( ) 函数，其效果如图 2-9 所示。

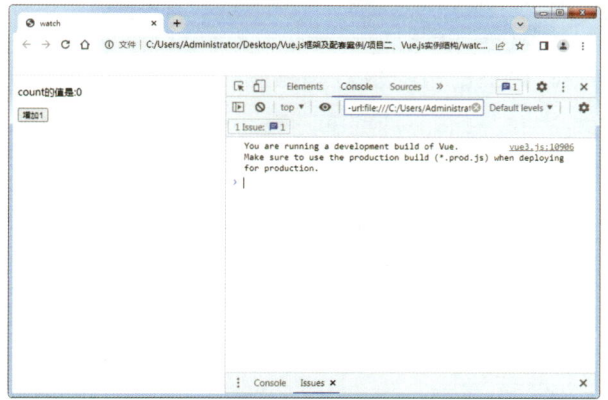

图 2-8 初始状态

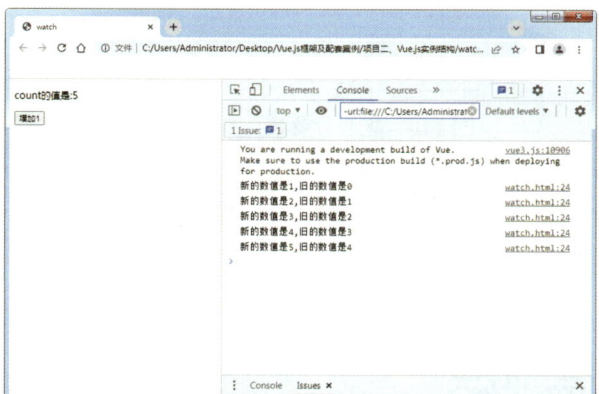

图 2-9 增加 count 数值

2.3 Vue 生命周期

2.3.1 Vue 生命周期的基本概念

Vue 生命周期是指 Vue 实例对象从创建之初到销毁之后的整个过程。Vue 所有功能的实现都是围绕其生命周期进行的，在生命周期的不同阶段调用对应的钩子函数来实现对应的功能。

Vue 生命周期可以分为八个阶段，分别是：beforeCreate（创建前）、created（创建后）、beforeMount（载入前）、mounted（载入后）、beforeUpdate（更新前）、updated（更新后）、beforeDestroy（销毁前）、destroyed（销毁后）。这里重点讲解一下前四个阶段。

创建前对应的钩子函数为 beforeCreate。此阶段初始化了一个 Vue 空的实例对象。在这个对象身上只有默认的一些生命周期函数和默认的事件，其他的东西都未创建。在这个阶段，data 和 methods 还没有初始化。

创建后对应的钩子函数为 created。在这个阶段，Vue 实例已经创建，但仍然不能获取 DOM 元素，而 data 和 methods 都已经初始化好了。如果要调用 methods 中的方法或者操作 data 中的数据，最早只能在 created 中操作。如果一个函数在打开页面就要执行，则需要在 created 中进行调用。

载入前对应的钩子函数是 beforeMount。在这一阶段，我们虽然依然得不到具体的 DOM 元素，但 Vue 挂载的根节点已经创建。

载入后对应的钩子函数是 mounted，这时候数据和 DOM 都已被渲染出来了。mounted 是平时使用较多的函数，网络的数据请求要写在这个阶段。

2.3.2 跑马灯效果

下面是一个制作跑马灯效果的案例。在 data 中定义变量 msg，使用 slice 函数将 msg 中的第一个字符和后面的字符截取出来后，将第一个字符放在最后面进行重新拼接，使用定时器将此操作每隔 500 毫秒执行一次。在 created 中调用 paomadeng() 函数，在打开浏览器的时候就能够执行 paomadeng() 函数了。

```html
<!DOCTYPE html>
<html lang="en">
<head>
    <meta charset="UTF-8">
    <meta http-equiv="X-UA-Compatible" content="IE=edge">
    <meta name="viewport" content="width=device-width, initial-scale=1.0">
    <title>跑马灯效果</title>
    <script src="./js/vue3.js"></script>
    <script>
      window.onload=function(){
        const app={
          data(){
            return {
              msg:'I like vue.js!',
            }
          },
          methods:{
            paomadeng(){
              setInterval(()=>{
                // 使用slice(0,1)截取msg中的第0个字符
                const start=this.msg.slice(0,1)
                // 使用slice(1,)截取第1个到最后的所有字符
                const end=this.msg.slice(1,)
                this.msg=end+start// 将start和end交换顺序后重新连接在一起
              },500)
            }


          },
          created(){
            this.paomadeng()
          }
        }
        Vue.createApp(app).mount('#container')
      }
```

> 定时器这里要写成箭头函数，否则的话 this 就获取不到该 Vue 实例

> 页面一加载就会执行 paomadeng() 函数

```
            </script>
        </head>
        <body>
            <div id="container">
                <h1>{{msg}}</h1>
            </div>
        </body>
</html>
```

在定时器 setInterval 中要使用箭头函数, 否则 this 的值就不再是该 Vue 实例。在 created 中调用 paomadeng( ) 函数时, 要在函数的前面添加 this。案例执行后的效果如图 2-10 和图 2-11 所示。

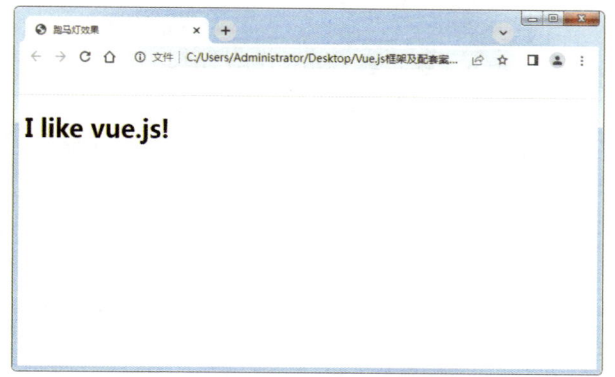

图 2-10　跑马灯效果 1

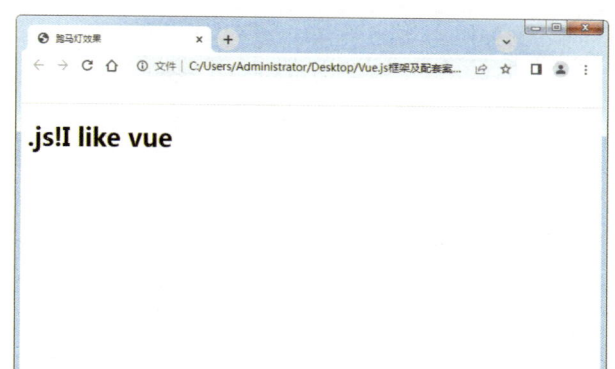

图 2-11　跑马灯效果 2

2.4　综合实训Ⅰ：制作计数器效果

综合实训Ⅰ：制作计数器效果

2.4.1　项目描述

在 data 中定义变量 n 的值为 0, 在 HTML 中将变量 n 渲染出来, 效果如图 2-12 所示。页面中有加号按钮和减号按钮, 点击加号让 n 的值增加, 到 5 以后数值不能再增大, 如图 2-13 所示。点击减号让 n 的值减少, 到 0 以后数值不能再减少, 如图 2-14 所示。

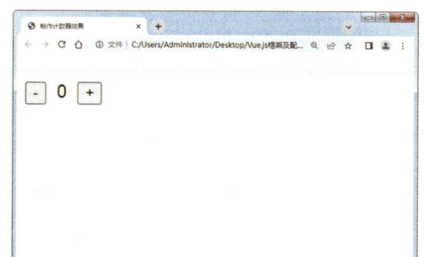
图 2-12　计数器初始状态

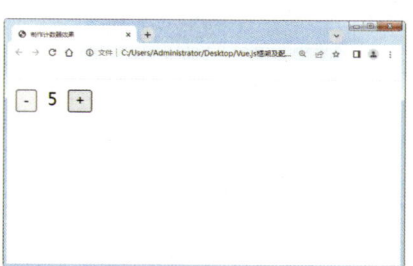
图 2-13　点击加号

图 2-14　点击减号

2.4.2　项目分析

计数器中的数字要先定义变量, 然后使用插值表达式进行数据绑定。在 methods 中定义两个方法,分别控制加号和减号按钮，同时使用 if 语句来控制数值增减的范围。

2.4.3 项目实施

定义变量 n 用于显示计数器中的数据，初始值定义为 0。定义两个函数，分别为 reduce( ) 和 add( )。reduce( ) 函数的功能是让 n 的数值减少，但是当减少到 0 以后不能再减少了。add( ) 函数的功能是让 n 的数值增加，当增加到 5 的时候不能再增加了。具体代码如下：

```html
<!DOCTYPE html>
<html lang="en">
<head>
    <meta charset="UTF-8">
    <meta http-equiv="X-UA-Compatible" content="IE=edge">
    <meta name="viewport" content="width=device-width, initial-scale=1.0">
    <title>计数器效果</title>
    <script src="./js/vue3.js"></script>
    <script>
        window.onload=function(){
            const app={

                data(){
                    return {
                        n:0              // 定义变量 n
                    }
                },
                methods:{
                    add(){               // add( ) 函数
                        if(this.n==5){
                            this.n=5
                        }
                        else{
                            this.n++
                        }
                    },
                    reduce(){            // reduce( ) 函数
                        if(this.n==0){
                            this.n=0
                        }
                        else{
                            this.n--
                        }
```

```
                    }
                }
            }
            Vue.createApp(app).mount('#container ')
        }
    </script>
</head>
<body>
    <div id="container" >
        <input type="button" value="-"  @click="reduce()">
        <span> {{n}} </span>
        <input type="button" value="+"  @click="add()">
    </div>

</body>
</html>
```

2.5 综合实训Ⅱ：制作逐字显示效果

综合实训Ⅱ：制作逐字显示效果

2.5.1 项目描述

在 data 中定义变量 str 的值为 "Vue 提供了一套声明式的、组件化的编程模型，帮助你高效地开发用户界面。"，在浏览器中运行代码后可以出现逐字显示的效果，如图 2-15 和图 2-16 所示。

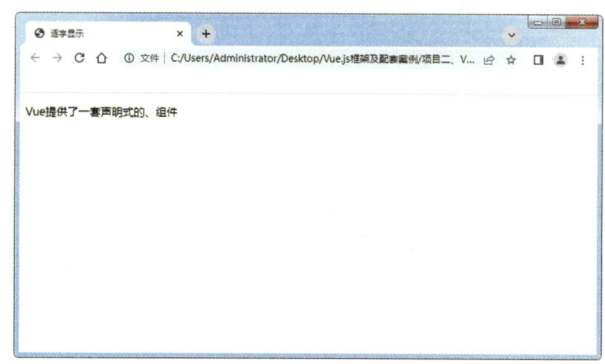

图 2-15　逐字显示效果 1

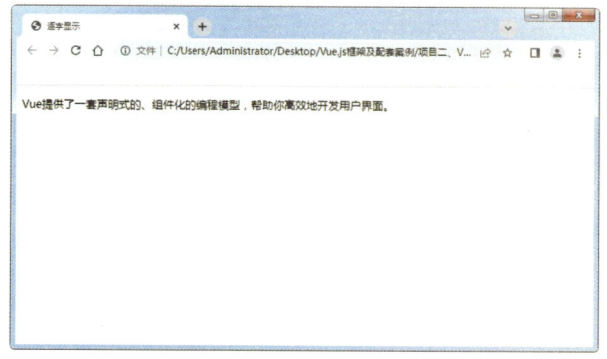

图 2-16　逐字显示效果 2

2.5.2 项目分析

首先使用变量定义好要显示的文字，使用插值表达式进行数据绑定。然后使用 slice( ) 函数对字符串进行截取，使用定时器每隔相同时间来执行一次 slice( ) 函数，让截取字符的数量每次增加一个，当字符串全部都出现后，关闭定时器。因为页面一加载就要执行，所以要使用 created。

2.5.3 项目实施

使用变量 str 定义一段字符，使用变量 i 记录要显示的文字的索引位置，使用变量 str2 来保存显示的文字，使用变量 timer 来命名定时器。如果 i 的值等于 str 字符串的长度，说明所有的文字都显示出来了，此时要关闭定时器，否则就让 i 的值增加 1。具体代码如下：

```html
<!DOCTYPE html>
<html lang="en">
<head>
    <meta charset="UTF-8">
    <meta http-equiv="X-UA-Compatible" content="IE=edge">
    <meta name="viewport" content="width=device-width, initial-scale=1.0">
    <title>逐字显示</title>
    <script src="./js/vue3.js"></script>
    <script>
        window.onload=function(){
            const app={
                data(){
                    return {
                        str:"Vue 提供了一套声明式的、组件化的编程模型，帮助你高效地开发用户界面。",
                        i:0,
                        str2:"",
                        timer:null
                    }
                },
                methods:{
                    showText(){
                        this.timer=setInterval(()=>{    // 使用箭头函数
                            if(this.i==this.str.length){
                                clearInterval(this.timer)
                            }
                            else{
                                this.i++
                            }
                            this.str2=this.str.slice(0,this.i)    // 截取 str 从 0 到 this.i 之间的字符串赋值给 str2
                        },200)
                    },
                    created(){
```

```
                    this.showText()
                }
            }
            Vue.createApp(app).mount("#container")
        }
    </script>
</head>
<body>
    <div id="container">
        <p>{{str2}}</p>   在页面中渲染变量 str2
    </div>
</body>
</html>
```

课后练习题

1. Vue 数据类型可以分为_____、_____、_____、_____、_____等。

2. 数据的绑定分为_____和_____，单向绑定使用插值表达式，符号为_____。

3. 在函数中，如果使用 data 中的变量则需要在变量前面添加_____，指代的是该 Vue 实例对象。

4. 计算属性使用_____来进行定义，它的本质是一个方法，在里面写一些计算逻辑的属性，然后使用 return 返回一个结果

5. watch 和 computed 两者中，_____不支持缓存，_____支持缓存。

6. 创建后对应的钩子函数为_____。在这个阶段，Vue 实例已经创建，但不能获取 DOM 元素，而 data 和 methods 都已经初始化好了

7. 载入后对应的钩子函数是_____，这时候数据和 DOM 都已被渲染出来了。网络的数据请求要写在这个阶段。

第 3 章

Vue 内置指令

导言

　　江河湖海日夜奔腾，是祖国汩汩的血液。五岳山川巍巍耸立，是祖国不屈的脊梁。万里长城绵延不绝，是祖国伟大与强盛的见证。每一帧祖国的美景都让人流连忘返。在本章中，我们将会学习 Vue 的内置指令。通过使用这些内置指令，我们可以制作图片切换的效果，让我们一起在案例中欣赏祖国的大好河山。

学习内容

　　本章主要讲解 Vue 的内置指令，一共有 6 节。3.1 节讲解了 v-text、v-cloak、v-html 指令。3.2 节讲解了 v-on 指令及修饰符。3.3 节讲解了 v-if、v-else、v-show 指令。3.4 节讲解了 v-for 指令。3.5 节讲解了 v-bind 指令及其综合应用案例。3.6 节讲解了 v-model 指令及其综合应用案例。

学习目标

1. 了解 Vue 的内置指令以及内置指令的使用方法。
2. 掌握 v-text、v-html 的使用及两者的区别。
3. 掌握 v-on 的使用方法以及键盘控制的方法。
4. 掌握 v-if、v-else、v-show 的使用方法。
5. 掌握 v-for、v-bind、v-model 的使用方法。

学习重点

1. 掌握 v-on 的使用方法。
2. 掌握 v-if、v-else、v-show 的使用方法。
3. 掌握 v-for、v-bind、v-model 的使用方法。

3.1 v-text、v-cloak、v-html 指令

3.1.1 v-text 指令

v-text 指令用于将变量显示为文本，它会替代原来标签中的文本。当绑定的数据对象上的值发生改变时，插值处的内容也会随之更新。下面的案例使用 v-text 将 msg 的数据渲染出来。

```
<!DOCTYPE html>
<html lang="en">
<head>
    <meta charset="UTF-8">
    <meta http-equiv="X-UA-Compatible" content="IE=edge">
    <meta name="viewport" content="width=device-width, initial-scale=1.0">
    <title>v-text 指令 </title>
    <script src="./js/vue3.js"></script>
    <script>
      window.onload=function(){
        const app={
          data(){
            return {
              msg:'hello vue!',
            }
          },

        }
        Vue.createApp(app).mount('#container')
      }
    </script>
</head>
<body>
    <div id="container">
        <p v-text="msg">原来的内容 </p>
    </div>
</body>
</html>
```

> p 标签原来的内容会被 msg 变量替换

3.1.2 v-cloak 指令

v-text 与插值表达式 {{}} 的作用是等同的。两者的区别是：在渲染的数据比较多的时候或者网络加载不好的情况下，插值表达式 {{}} 可能会先把大括号显示出来，等数据全部加载出来以后才正常显示。

解决的方式是使用 v-cloak。

```html
<!DOCTYPE html>
<html lang="en">
<head>
    <meta charset="UTF-8">
    <meta http-equiv="X-UA-Compatible" content="IE=edge">
    <meta name="viewport" content="width=device-width, initial-scale=1.0">
    <title> v-cloak 指令 </title>
    <style>
        [v-cloak]{          // 在样式中设置 [v-cloak]
            display: none !important;
        }
    </style>
    <script src="./js/vue3.js"></script>
    <script>
      window.onload=function(){
        const app={
          data(){
            return {
              msg:'hello vue!',
            }
          },

        }
        Vue.createApp(app).mount('#container')
      }
    </script>
</head>
<body>
    <div id="container" v-cloak>          // 在标签中添加 v-cloak
      <p v-text="msg"> 原来的内容 </p>
      <p>{{msg}}</p>
    </div>
</body>
</html>
```

v-cloak 指令的作用是通过在样式中给属性 v-cloak 添加样式，让它一开始显示为"无"，即是隐藏的状态，当关联的 Vue 实例完全加载后会删除 v-cloak 属性。

3.1.3　v-html 指令

v-html 指令与 v-text 指令相似，区别是 v-html 指令能够解析变量中的 HTML 标签，但是 v-text 会将变量全部输出为文本。下面案例对比了 v-text 和 v-html 两个指令的输出结果。

```html
<!DOCTYPE html>
<html lang="en">
<head>
    <meta charset="UTF-8">
    <meta http-equiv="X-UA-Compatible" content="IE=edge">
    <meta name="viewport" content="width=device-width, initial-scale=1.0">
    <title>v-html 指令</title>
    <script src="./js/vue3.js"></script>
    <script>
      window.onload=function(){
        const app={
          data(){
            return {
                msg:'<a href="http://www.baidu.com">百度</a>',
            }
          },
        }
        Vue.createApp(app).mount('#container')
      }
    </script>
</head>
<body>
    <div id="container" >
      <p v-text="msg"></p>
      <p v-html="msg"></p>
    </div>
</body>
</html>
```

案例执行后的效果如图 3-1 所示。

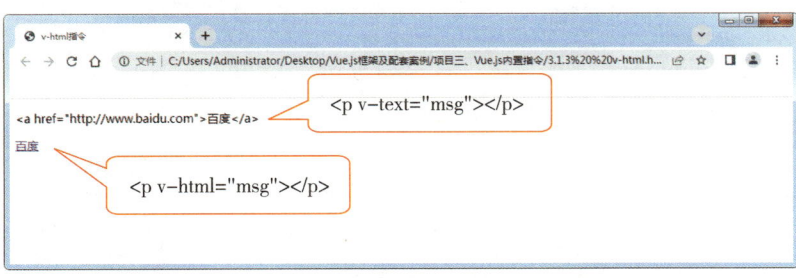

图 3-1　v-text 和 v-html 指令的输出结果对比

3.2 v-on 指令

3.2.1 v-on 基本用法

v-on 用来给元素添加事件。v-on 的语法为 v-on: 事件名 =" 事件函数 "，例如 v-on:click="add( )" 的含义是当触发了点击事件以后会执行 add( ) 函数。v-on 可以简写为 @，上面的 v-on:click="add( )" 可以改写为 @click="add( )"。下面案例点击按钮"照镜子"可以实现文字翻转效果。

```html
<!DOCTYPE html>
<html lang="en">
<head>
    <meta charset="UTF-8">
    <meta http-equiv="X-UA-Compatible" content="IE=edge">
    <meta name="viewport" content="width=device-width, initial-scale=1.0">
    <title>v-for 基本用法 </title>
    <script src="./js/vue3.js"></script>
    <script>
      window.onload=function(){
        const app={
          data(){
            return {
              str:"Vue",
              str2:""
            }
          },
          methods:{
            fanzhuan(){
              this.str2=this.str.split("").reverse().join("")
            }
          }
        }
        Vue.createApp(app).mount('#container')
      }
    </script>
</head>
<body>
    <div id="container">
      <input type="button" value=" 照镜子 " @click="fanzhuan()">
      <p><span>{{str}}</span> | <span>{{str2}}</span></p>
```

```
        </div>
    </body>
</html>
```

点击"照镜子"按钮之前的效果如图 3-2 所示，点击按钮以后的效果如图 3-3 所示。

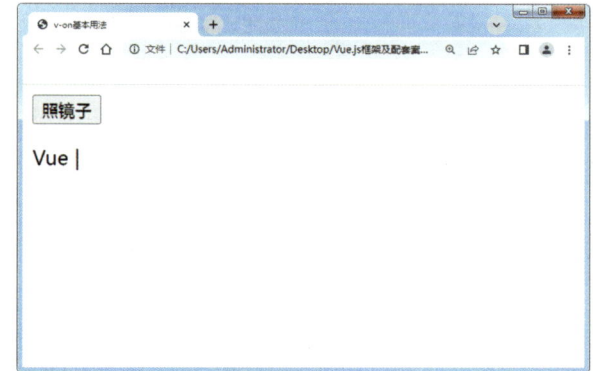

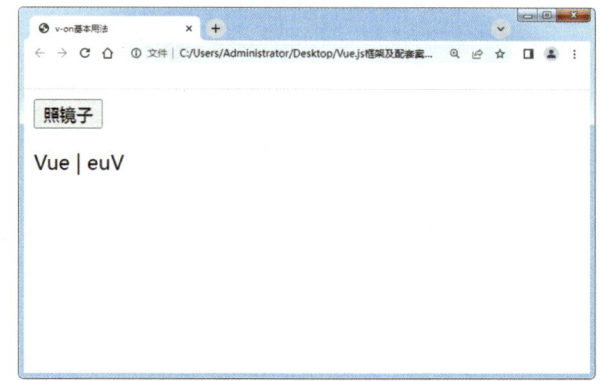

图 3-2　点击按钮之前

图 3-3　点击按钮之后

3.2.2　事件修饰符

事件修饰符是在事件后面添加"."开头的指令后缀。常用的事件修饰符有 prevent、stop、once 等。其中 prevent 的作用是阻止默认事件，stop 的作用是阻止事件冒泡，once 的作用是只触发一次事件。

```
<!DOCTYPE html>
<html lang="en">
<head>
    <meta charset="UTF-8">
    <meta http-equiv="X-UA-Compatible" content="IE=edge">
    <meta name="viewport" content="width=device-width, initial-scale=1.0">
    <title>事件修饰符</title>
    <style>
        .box{ width:300px;height:100px;padding:20px;background-color:darkseagreen;}
        .box p{ width:200px;height:50px;background-color:lightpink;border:2px#000000 solid;}
    </style>
    <script src="./js/vue3.js"></script>
    <script>
      window.onload=function(){
        const app={
          methods:{
            showInfo1(){
              alert("阻止默认事件")
            },
```

```
            showInfo2(){
                alert("父对象")
            },

            showInfo3(){
                alert("子对象")
            },
            showInfo4(){
                alert("只执行一次")
            }
          }
        }
        Vue.createApp(app).mount('#container')
      }
    </script>
</head>
<body>
    <div id="container">
        <h4>1.阻止默认事件使用prevent</h4>
        <a href="http://www.baidu.com" @click.prevent="showInfo1()">百度</a>
        <hr>
        <h4>2.阻止事件冒泡使用stop</h4>
        <div class="box" @click="showInfo2()">
           父对象
           <p @click.stop="showInfo3()">子对象</p>
        </div>
        <hr>
        <h4>3.只执行一次使用once</h4>
        <input type="button" value="点击按钮" @click.once="showInfo4()">
    </div>
</body>
</html>
```

`<a href="http://www.baidu.com" @click.prevent="showInfo1()">百度</a>`

点击"百度"两个字，默认情况下要进行链接跳转，添加了prevent以后会阻止默认事件，即不会进行链接跳转，效果如图3-4所示。

`<p @click.stop="showInfo3()">子对象</p>`

p标签是 `<div class="box"></div>` 的子标签，点击div会触发showInfo2()函数，点击p标签会触发showInfo3()，同时也会触发父标签div中的showInfo2()函数，给p标签添加了stop以后会阻止事件冒泡，即不再执行父一级标签上面的事件，效果如图3-5所示。

```
<input type="button" value="点击按钮" @click.once="showInfo4()">
```

once 的作用是让事件只执行一次。点击按钮第一次的时候会执行 showInfo4() 函数，但是再点击就不再执行了，效果如图 3-6 所示。

图 3-4 阻止默认事件

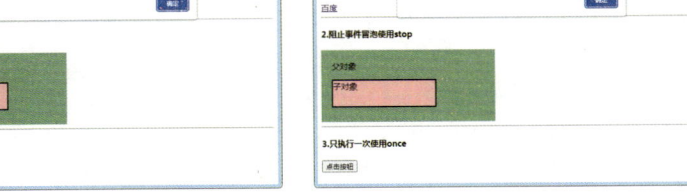
图 3-5 阻止事件冒泡

图 3-6 只执行一次

3.2.3 常用事件类型

常用的事件有：click（鼠标点击）、mouseover（鼠标经过）、mouseout（鼠标离开）、keydown（键盘按下）、keyup（键盘弹起）、input（文本框输入）、focus（元素获取焦点）、blur（元素失去焦点）。下面案例使用了键盘事件。

案例中有一个变量 num，初始值为 5。在文本框中按上箭头键，num 的值会增大，按下箭头键，num 的值会减少，按回车键会弹出一个警示框。具体代码如下：

```html
<!DOCTYPE html>
<html lang="en">
<head>
    <meta charset="UTF-8">
    <meta http-equiv="X-UA-Compatible" content="IE=edge">
    <meta name="viewport" content="width=device-width, initial-scale=1.0">
    <title>键盘事件</title>
    <style>
        .box{ width:300px;height:100px;padding:20px;background-color:darkseagreen;}
        .box p{ width:200px;height:50px;background-color:lightpink;border:2px #000000 solid;}
    </style>
    <script src="./js/vue3.js"></script>
    <script>
      window.onload=function(){
```

```
const app={
    data(){
        return {
            num:5
        }
    },
    methods:{
        show(){
            alert('num 的值为 :'+this.num)
        }
    }
}
Vue.createApp(app).mount('#container')
</script>
</head>
<body>
    <div id="container">
        <p>num 的值是 :{{num}}</p>
        <input type="text" @keyup.up="num++" @keyup.down="num--" @keyup.enter="show()">
    </div>
</body>
</html>
```

按下键盘上的上箭头键后的效果如图 3-7 所示，按下箭头后的效果如图 3-8 所示，按回车键后的效果如图 3-9 所示。

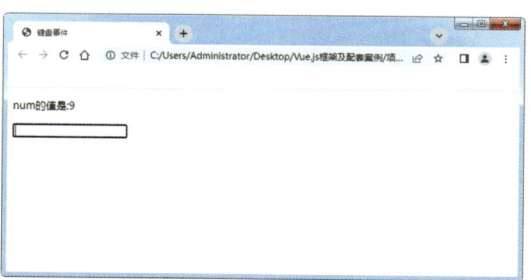

图 3-7　按上箭头

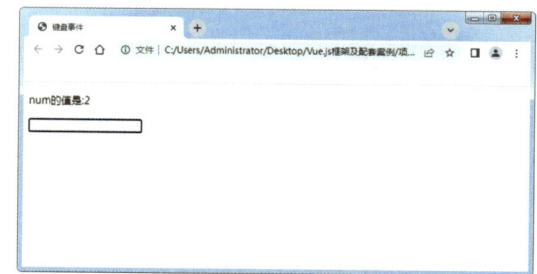

图 3-8　按下箭头

图 3-9　按回车键

3.3 v-if、v-else、v-show 指令

3.3.1 v-if 指令

v-if 指令用于对象的显示和隐藏。当 v-if 的值为 true（真）时，对象显示；当 v-if 的值为 false（假）时，对象隐藏。下面案例通过点击按钮切换变量 flag 的值来改变 div 的显示和隐藏。

```
<!DOCTYPE html>
<html lang="en">
<head>
    <meta charset="UTF-8">
    <meta http-equiv="X-UA-Compatible" content="IE=edge">
    <meta name="viewport" content="width=device-width, initial-scale=1.0">
    <title>v-if 指令</title>
    <script src="./js/vue3.js"></script>
    <script>
      window.onload=function(){
        const app={
          data(){
            return {
                flag:true,
            }
          },
          methods:{
            change(){
              this.flag=!this.flag// 作用是让 flag 的值取反
            }
          }
        }
        Vue.createApp(app).mount('#container')
      }
    </script>
</head>
<body>
    <div id="container">
      <input type="button" value="点击切换图片显示" @click="change()">
      <div class="pic">
        <img src="./images/changcheng.jpg" alt="长城图片" v-if="flag">
      </div>
```

```
        </div>
    </body>
</html>
```

案例效果为：刚开始时，图片显示，如图 3-10 所示；当点击按钮后，图片会隐藏，如图 3-11 所示，再点击按钮，图片又显示。

图 3-10　图片显示

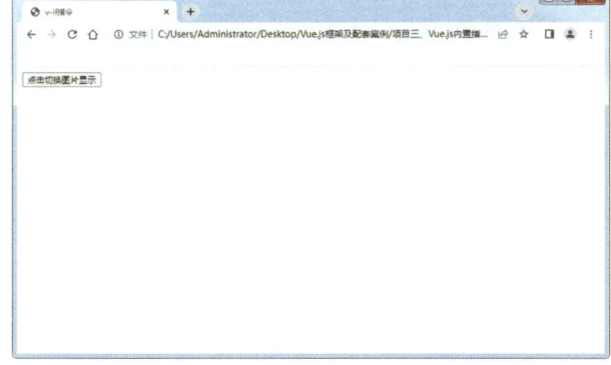

图 3-11　图片隐藏

该案例也可以不使用 change 函数，直接将 flag 取反的代码写在标签内部。具体代码如下：

```
<!DOCTYPE html>
<html lang="en">
<head>
    <meta charset="UTF-8">
    <meta http-equiv="X-UA-Compatible" content="IE=edge">
    <meta name="viewport" content="width=device-width, initial-scale=1.0">
    <title>v-if 指令</title>
    <script src="./js/vue3.js"></script>
    <script>
      window.onload=function(){
        const app={
          data(){
            return {
              flag:true,
            }
          },
        }
        Vue.createApp(app).mount('#container')
      }
    </script>
</head>
<body>
    <div id="container">
```

```
        <input type="button" value="点击切换图片显示" @click="flag=!flag">
        <div class="pic">
          <img src="./images/changcheng.jpg" alt="长城图片 " v-if="flag">
        </div>
    </div>
</body>
</html>
```

> 这里变量 flag 前面不能加 this，只有 Vue 实例中才使用 this

3.3.2 v-else 指令

v-else 指令要结合 v-if 使用。当 v-if 的值为真时，v-else 的值为假；当 v-if 的值为假时，v-else 的值为真。下面案例通过点击按钮在两张图之间切换显示。

```
<!DOCTYPE html>
<html lang="en">
<head>
    <meta charset="UTF-8">
    <meta http-equiv="X-UA-Compatible" content="IE=edge">
    <meta name="viewport" content="width=device-width, initial-scale=1.0">
    <title>v-else 指令 </title>
    <script src="./js/vue3.js"></script>
    <script>
      window.onload=function(){
        const app={
          data(){
            return {
              flag:true,
            }
          },
        }
        Vue.createApp(app).mount('#container')
      }
    </script>
</head>
<body>
    <div id="container">
        <input type="button" value="点击两张图片切换" @click="flag=!flag">
        <div class="pic" >
          <img src="./images/changcheng.jpg" alt="长城图片1" v-if="flag">
          <img src="./images/changcheng2.jpg" alt="长城图片2" v-else>
        </div>
```

```
        </div>
    </body>
</html>
```

当初始 flag 的值为 true 时，长城图片 1 显示，长城图片 2 隐藏，如图 3-12 所示；点击按钮以后，flag 的值变成 false，长城图片 1 隐藏，长城图片 2 显示，如图 3-13 所示。

图 3-12　长城图片 1 显示

图 3-13　长城图片 2 显示

3.3.3　v-show 指令

v-show 指令与 v-if 指令类同，都是使用真或者假来控制元素的显示或隐藏。区别在于 v-if 是从 DOM 中添加或删除元素，而 v-show 不是从 DOM 中删除元素，而是使用 CSS 的 display 属性来控制元素的显示或隐藏。下面的案例对比了 v-if 和 v-show 的区别。

```
<!DOCTYPE html>
<html lang="en">
<head>
    <meta charset="UTF-8">
    <meta http-equiv="X-UA-Compatible" content="IE=edge">
    <meta name="viewport" content="width=device-width, initial-scale=1.0">
    <title>v-show 指令</title>
    <script src="./js/vue3.js"></script>
    <script>
      window.onload=function(){
        const app={
          data(){
            return {
                flag:true,
            }
          },
        }
        Vue.createApp(app).mount('#container')
```

```html
        }
    </script>
</head>
<body>
    <div id="container">
        <input type="button" value="点击切换图片显示" @click="flag=!flag">
        <div class="pic">
            <h4>使用v-if来控制图片的显示和隐藏</h4>
            <img src="./images/changcheng.jpg" alt="长城图片" width="300" v-if="flag">
            <hr>
            <h4>使用v-show来控制图片的显示和隐藏</h4>
            <img src="./images/changcheng.jpg" alt="长城图片" width="300" v-show="flag">
        </div>
    </div>
</body>
</html>
```

执行代码的时候因为flag的值为true，所以两张图片都会显示，如图3-14所示；当点击按钮的时候，两张图片都消失，如图3-15所示。使用Chrome浏览器，按F12键进入控制台可以看到使用v-if消失的图片从DOM结构中删除了，使用v-show消失的图片在DOM结构中依然存在，只是样式变成了display:none。

图3-14 两张图片都显示

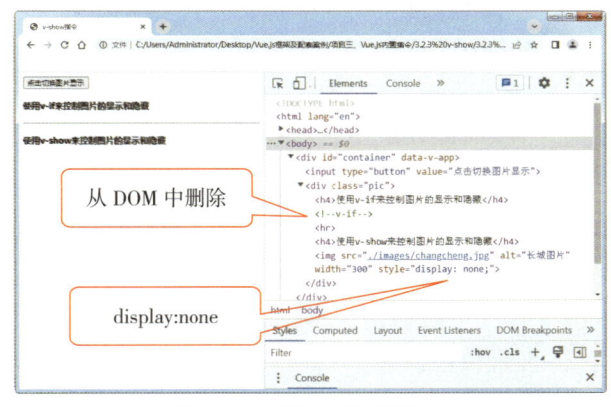

图3-15 两张图片隐藏

在不同的使用场景中，对于频繁切换的元素优先选择v-show。因为使用v-show，元素的渲染不需要重复进行，而v-if需要重新渲染元素及其所有子元素，因此对于频繁切换的元素具有较高的开销。

3.4 v-for指令

3.4.1 v-for基本用法

v-for是Vue的一个核心指令，用于循环可迭代的数据，例如数组和对象等。下面案例通过使用

v-for 来循环显示一个数组。

```html
<!DOCTYPE html>
<html lang="en">
<head>
    <meta charset="UTF-8">
    <meta http-equiv="X-UA-Compatible" content="IE=edge">
    <meta name="viewport" content="width=device-width, initial-scale=1.0">
    <title>v-for基本用法</title>
    <script src="./js/vue3.js"></script>
    <script>
      window.onload=function(){
        const app={
          data(){
            return {
                fruit:["苹果","香蕉","西瓜","草莓"]
            }
          },
        }
        Vue.createApp(app).mount('#container')
      }
    </script>
</head>
<body>
    <div id="container">
      <ul>
        <li v-for="(value,index) in fruit" :key="index">
          {{index}}--{{value}}
        </li>
      </ul>
    </div>
</body>
</html>
```

执行代码后的效果如图 3-16 所示。

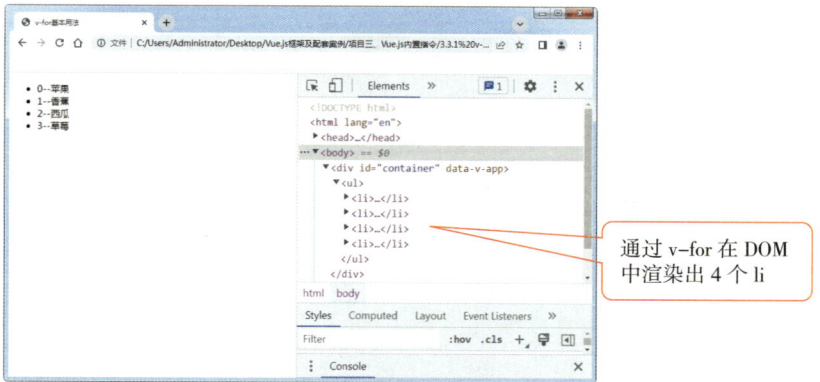

图 3-16　v-for 的基本用法

```
<li v-for="(value,index) in fruit" :key="index">
```

其中，fruit 是数组名称；in 是 v-for 中的固定结构；value 和 index 是两个变量名，可以换成其他的变量名称，（value,index）中前面一个变量的作用是获取数据的值，后一个变量的作用是获取数据在集合中的排序；:key 属性用于标识遍历出来的元素，:key 的值只能是 number 或者 string 类型的数据。

3.4.2　使用 v-for 遍历复杂数据

当数据比较复杂的时候，使用 v-for 的方法是相同的。下面案例演示了使用 v-for 来遍历出一个数组内部的三个结构复杂的数据。

```
<!DOCTYPE html>
<html lang="en">
<head>
    <meta charset="UTF-8">
    <meta http-equiv="X-UA-Compatible" content="IE=edge">
    <meta name="viewport" content="width=device-width, initial-scale=1.0">
    <title>v-for基本用法</title>
    <script src="./js/vue3.js"></script>
    <script>
      window.onload=function(){
        const app={
          data(){
            return {
              chengji:[
                {name:"小明",class:"1班",score:{
                  Chinese:99,
                  Math:96,
                  English:93
                }},

                {name:"小军",class:"2班",score:{
```

```
                    Chinese:85,
                    Math:99,
                    English:91
                }},
                {name:"小红",class:"3班",score:{
                    Chinese:98,
                    Math:96,
                    English:97
                }},

            ]
        }
    },
}
Vue.createApp(app).mount('#container')
        }
    </script>
</head>
<body>
    <div id="container">
        <table width="600" border="1">
            <tr>
                <th>姓名</th>
                <th>班级</th>
                <th>成绩</th>
            </tr>
            <tr v-for="(value,index) in chengji" :key="index" align="center">
                <td>{{value.name}}</td>
                <td>{{value.class}}</td>
                <td>
                    <p>语文:{{value.score.Chinese}}</p>
                    <p>数学:{{value.score.Math}}</p>
                    <p>英语:{{value.score.English}}</p>
                </td>
            </tr>
        </table>
    </div>
</body>
</html>
```

执行代码后的效果如图 3-17 所示。

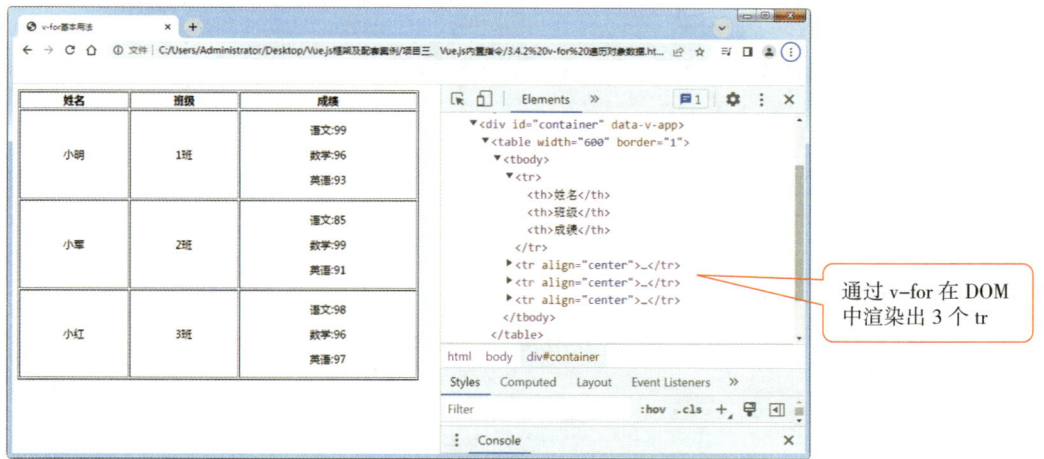

图 3-17 循环复杂数据

3.5 v-bind 指令

3.5.1 v-bind 属性绑定

标签属性是对标签的一种描述方式，通常由"属性名 = 属性值"组成，例如下面的 img 标签有三个属性，分别为 src、alt、width。

```
<img src="./images/changcheng.jpg" alt="长城图片" width="500" >
```

属性绑定的含义是将原来固定的属性值改为使用 data 中的变量，属性绑定使用 v-bind 指令。具体代码如下：

```
<!DOCTYPE html>
<html lang="en">
<head>
    <meta charset="UTF-8">
    <meta http-equiv="X-UA-Compatible" content="IE=edge">
    <meta name="viewport" content="width=device-width, initial-scale=1.0">
    <title>v-bind 属性绑定 </title>
    <script src="./js/vue3.js"></script>
    <script>
      window.onload=function(){
        const app={
          data(){
            return {
              imgSrc:"./images/changcheng.jpg",
              imgW:500
            }
```

```
            },
        }
        Vue.createApp(app).mount('#container')
    }
    </script>
</head>
<body>
    <div id="container">
        <img v-bind:src="imgSrc" alt="长城图片" v-bind:width="imgW" >
    </div>
</body>
</html>
```

"v-bind: 属性"可以简写为":属性",例如 v-bind:src="imgSrc" 可以简写为 :src="imgSrc",即上面案例中的 \<img v-bind:src="imgSrc"alt=" 长城图片 "v-bind:width="imgW"\> 可以改写为：

```
<img :src="imgSrc" alt="长城图片" :width="imgW" >
```

3.5.2 使用 v-bind 实现图片切换

使用 v-bind 实现图片切换的思路是设置一个数组,在数组中放置三张图片的路径,然后通过更改数组的编号来实现图片的切换。案例代码如下：

```
<!DOCTYPE html>
<html lang="en">
<head>
    <meta charset="UTF-8">
    <meta http-equiv="X-UA-Compatible" content="IE=edge">
    <meta name="viewport" content="width=device-width, initial-scale=1.0">
    <title>使用 v-bind 实现图片切换</title>
    <style>
      *{ margin: 0; padding: 0;}
      .pic{ width: 600px; height: 300px; margin: 100px auto; border:2px red solid; position: relative;}
      .pic img{ width: 600px; height: 300px;}
        .pic .left,.pic .right{ width: 40px; height: 60px; background-color: rgba(0,0,0,0.5); text-align: center; line-height: 60px; position: absolute; top:120px; color: #FFF; font-size: 40px; cursor: pointer;}
        .pic .left{ left: 0;}
        .pic .right{ right: 0;}
    </style>
    <script src="./js/vue3.js"></script>
    <script>
```

使用 v-bind 实现图片切换

```
window.onload=function(){
  const app={
    data(){
      return {
        imgSrc:[
          './images/pic1.jpg',
          './images/pic2.jpg',
          './images/pic3.jpg',
          './images/pic4.jpg'
        ],
        i:0,
      }
    },
    methods:{
      prevPic(){
        if(this.i==0){
          this.i=this.imgSrc.length-1
        }
        else{
          this.i--
        }
      },
      nextPic(){
        if(this.i==this.imgSrc.length-1){
          this.i=0
        }
        else{
          this.i++
        }
      },
    }
  }
  Vue.createApp(app).mount('#container')
}
</script>
</head>
<body>
  <div id="container">
    <div class="pic">
```

```
            <img :src="imgSrc[i]" alt="">
            <div class="left" @click="prevPic">&lt;</div>
            <div class="right" @click="nextPic">&gt;</div>
        </div>
    </div>
</body>
</html>
```

页面加载后的效果如图 3-18 所示，prevPic 函数的功能是让 i 的值减少 1，即图片向前一张。如果 i 的值减少到 0，则将 i 的值变为 this.imgSrc.length-1，即最后一张图的编号。nextPic 函数的功能是让 i 的值增加 1，即图片向后一张。如果 i 的值增大到 this.imgSrc.length-1，即最后一张图的编号，则将 i 的值变为 0，切换效果如图 3-19 所示。

图 3-18　图片切换 1

图 3-19　图片切换 2

3.5.3　图片自动切换

该案例是在上一个案例的基础上添加了定时器，实现了打开页面图片就自动切换的效果。当鼠标经过图片 div 的时候定时器关闭，图片停止自动切换，当鼠标离开的时候图片再次开始自动切换。案例代码如下：

```
<!DOCTYPE html>
<html lang="en">
<head>
    <meta charset="UTF-8">
    <meta http-equiv="X-UA-Compatible" content="IE=edge">
    <meta name="viewport" content="width=device-width, initial-scale=1.0">
    <title>添加定时器实现图片自动切换</title>
    <style>
        *{ margin: 0; padding: 0;}
        .pic{ width: 600px; height: 300px;  margin: 100px auto; border:2px red solid; position: relative;}
        .pic img{ width: 600px; height: 300px;}
```

```
            .pic .left,.pic .right{ width: 40px; height: 60px; background-color: rgba(0,0,0,0.5);
text-align: center;  line-height: 60px; position: absolute; top:120px; color: #FFF; font-size:
40px; cursor: pointer;}
            .pic .left{ left: 0;}
            .pic .right{ right: 0;}
    </style>
    <script src="./js/vue3.js"></script>
    <script>
        window.onload=function(){
            const app={
                data(){
                    return {
                        imgSrc:[
                            './images/pic1.jpg',
                            './images/pic2.jpg',
                            './images/pic3.jpg',
                            './images/pic4.jpg'
                        ],
                        i:0,
                        timer:null,
                    }
                },
                methods:{
                    prevPic(){
                        if(this.i==0){
                            this.i=this.imgSrc.length-1
                        }
                        else{
                            this.i--
                        }
                    },

                    nextPic(){
                        if(this.i==this.imgSrc.length-1){
                            this.i=0
                        }
                        else{
                            this.i++
                        }
                    },
```

```
          autoChange(){
            this.timer=setInterval(()=>{
              this.nextPic()
            },1000)
          },

          stopChange(){
            clearInterval(this.timer)
          }
        },
        created(){
          this.autoChange()
        },
      }
      Vue.createApp(app).mount('#container')
    }
    </script>
  </head>
  <body>
    <div id="container">
      <div class="pic" @mouseover="stopChange()" @mouseout="autoChange()">
        <img :src="imgSrc[i]" alt="">
        <div class="left" @click="prevPic">&lt;</div>
        <div class="right" @click="nextPic">&gt;</div>
      </div>
    </div>
  </body>
</html>
```

该案例中的 autoChange( ) 函数使用了定时器，让图片每隔 1 000 毫秒，即 1 秒的间隔就切换一张图片，同时在页面加载的时候，即生命周期的 created 中执行 autoChange( ) 函数，从而实现了页面加载以后图片能自动切换的效果。

stopChange( ) 函数的作用是停止定时器。鼠标经过类 pic 的 div 时，执行 stopChange( ) 函数，图片停止切换，鼠标离开 div 的时候执行 autoChange( ) 函数，图片自动切换。

3.5.4 类和行间样式的绑定

类的绑定，即类名使用的是一个动态改变的值。类的绑定格式是 :class="{ 类名称 : 布尔值 }"，如果布尔值为真则使用该类，例如当代码为 :class="{bd1:true}" 时会添加类 bd1；如果布尔值为假，则不使用该类，例如当代码变为 :class="{bd1:false}" 时就不会添加类 bd1。

行间样式绑定的格式为 :style="{ 样式 : 变量 }"，例如 :style="{'background-image':bg}"，其中 bg 是在 data 中定义的一个变量。下面案例演示了类和行间样式的绑定，具体代码如下：

```html
<!DOCTYPE html>
<html lang="en">
<head>
    <meta charset="UTF-8">
    <meta http-equiv="X-UA-Compatible" content="IE=edge">
    <meta name="viewport" content="width=device-width, initial-scale=1.0">
    <title> 类和行间样式的绑定 </title>
    <style>
      .box1{ width: 500px; height: 400px ;background-color: lightgreen;}
      .bd1{ border:5px red solid }
      .bd2{ border:5px red dotted}
    </style>
    <script src="./js/vue3.js"></script>
    <script>
      window.onload=function(){
        const app={
          data(){
            return {
                flag:true,
                bgSrc:["url(./images/bg1.png)","url(./images/bg2.png)"],
                i:0,
            }
          },
        }
        Vue.createApp(app).mount('#container')
      }
    </script>
</head>
<body>
    <div id="container">
      <input type="button" value="切换类" @click="flag=!flag">
      <div class="box1" :class="{bd1:flag,bd2:!flag}"></div>
      <hr>
      <input type="button" value="切换行间样式" @click="i=1">
      <div class="box1" :style="{'background-image':bgSrc[i]}"></div>
    </div>
</body>
</html>
```

> 点击按钮后，flag 的值取反

> 样式中间有连接符"-"，所以样式的外面必须有引号

案例中样式 .bd1 为 5px 的红色实线，样式 .bd2 为 5px 的红色虚线，变量 flag 的初始值为 true，所以执行 :class="{bd1:flag,bd2:!flag}" 对应的代码为 :class="{bd1:true,bd2:false}"。页面加载后，div 有红色的实线边框，效果如图 3-20 所示。当点击"切换类"按钮时，flag 取反，变成 false，这时 div 有红色的虚线边框，效果如图 3-21 所示。

变量 bgSrc 是一个数组，里面有两个值，i 的初始值为 0，所以页面加载后，div 的背景图为 bg1.png，效果如图 3-22 所示。点击按钮"切换行间样式"后，i 的值变成 1，这时背景图变为 bg2.png，效果如图 3-23 所示。

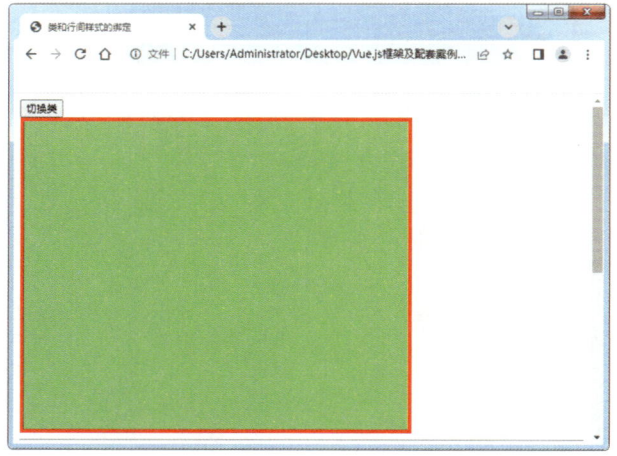

图 3-20　红色实线边框

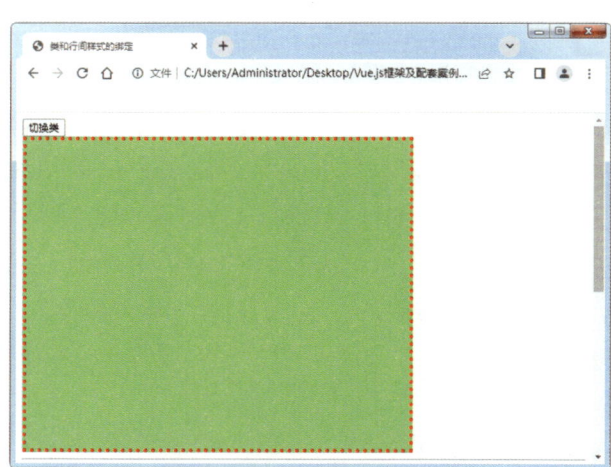

图 3-21　红色虚线边框

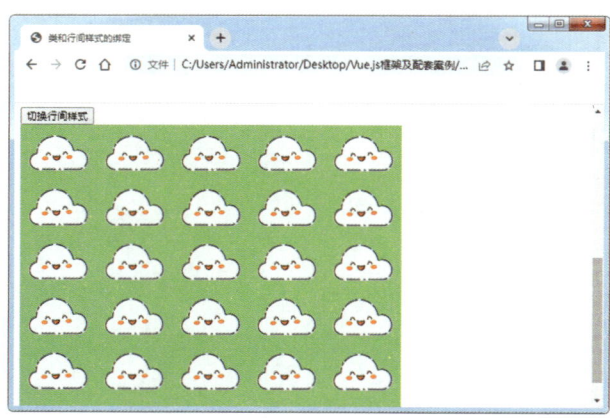

图 3-22　bg1 背景图

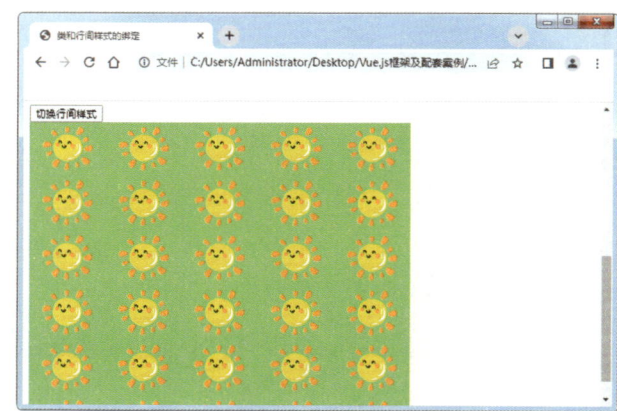

图 3-23　bg2 背景图

3.6　v-model 指令

3.6.1　v-model 指令基本使用

Vue 的核心特点之一是数据的响应式更新，v-model 是 Vue 实现这一特点的重要指令。v-model 指令可以实现数据的双向绑定，即 data 中的数据会在视图中显示，同时在视图中进行数据改变时也可以使 data 中的数据同步修改。

v-model 指令应用在表单输入元素上，包括输入框（如 text、password 等）、单选按钮（如 radio）、复选框（如 checkbox）和下拉框（如 select）等。

下面案例为文本框双向绑定了 data 中的变量 num，在 p 标签中通过 {{num}} 可以直观地看到 num 的值，因为 num 的值为 0，文本框将变量 data 的值显示出来，效果如图 3-24 所示。修改文本框的值，由于是双向绑定，data 的值相应地也被修改，此时 num 的值可以在下面的 p 标签中直观地显示出来，效果如图 3-25 所示。

```
<!DOCTYPE html>
<html lang="en">
<head>
    <meta charset="UTF-8">
    <meta http-equiv="X-UA-Compatible" content="IE=edge">
    <meta name="viewport" content="width=device-width, initial-scale=1.0">
    <title>v-model 双向数据绑定</title>
    <script src="./js/vue3.js"></script>
    <script>
      window.onload=function(){
        const app={
          data(){
            return {
              num:0
            }
          },
        }
        Vue.createApp(app).mount('#container')
      }
    </script>
</head>
<body>
    <div id="container">
        <input type="text" v-model="num">        双向绑定 data 中
        <hr>                                      的 num 变量
        <p>num 目前的值是：{{num}}</p>
    </div>                                        直观显示出 num
</body>                                           的值
</html>
```

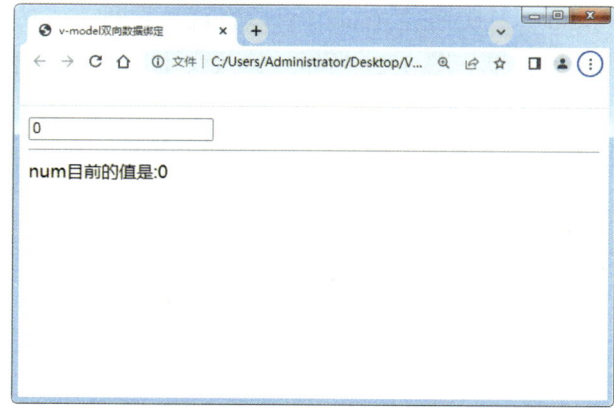
图 3-24　v-model 数据双向绑定 1

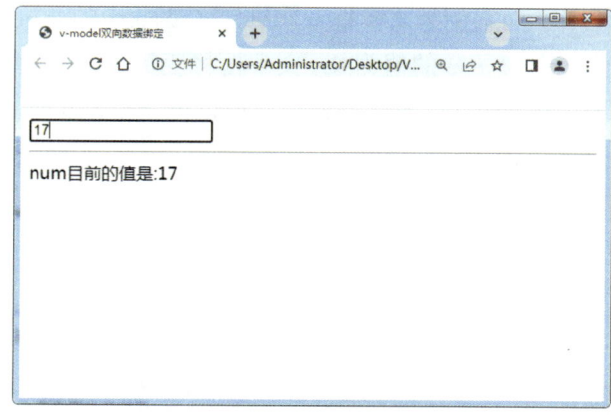
图 3-25　v-model 数据双向绑定 2

3.6.2　多种表单元素使用 v-model

除了文本框可以使用 v-model 来实现数据的双向绑定之外，表单中的其他元素，如单选按钮、复选框、下拉菜单等都可以使用 v-model 实现数据的双向绑定。下面案例对多种表单元素的双向数据绑定进行了演示。

```
<!DOCTYPE html>
<html lang="en">
<head>
    <meta charset="UTF-8">
    <meta http-equiv="X-UA-Compatible" content="IE=edge">
    <meta name="viewport" content="width=device-width, initial-scale=1.0">
    <title>多种表单对象中 v-model 的应用</title>
    <script src="./js/vue3.js"></script>
    <script>
      window.onload=function(){
        const app={
          data(){
            return {
              name:"",
              gender:"",
              hobbies:[],
              city:""
            }
          },
        }
        Vue.createApp(app).mount('#container')
      }
    </script>
</head>
<body>
```

多种表单对象中 v-model 的应用

```html
<div id="container">
    <p>用户名:<input type="text" v-model="name"></p>
    <p>性别:
        <input type="radio" name="sex" value="male" v-model="gender"> 男
        <input type="radio" name="sex" value="female" v-model="gender"> 女
    </p>
    <p>
        爱好:
        <input type="checkbox" value="read" v-model="hobbies"> 读书
        <input type="checkbox" value="code" v-model="hobbies"> 编程
        <input type="checkbox" value="sports" v-model="hobbies"> 运动
    </p>
    <p>
        城市:
        <select v-model="city">
            <option value="beijing">北京</option>
            <option value="shanghai">上海</option>
            <option value="guangzhou">广州</option>
        </select>
    </p>
    <hr>
    {{name}}你好,你的性别是{{gender}},你的爱好有:{{hobbies.join(",")}},你来自的城市
    是:{{city}}
</div>
</body>
</html>
```

> join 的作用是使用逗号(,)将数组中的数据连接成字符串

多个单选按钮以组的形式与 data 中的变量 gender 进行了双向数据绑定,当单选按钮中选择的值进行改变的时候,gender 的值就进行改变,gender 获取的值是被选中的单选按钮对应的 value 值。复选框的使用方法与单选按钮相似。下拉菜单是在 select 标签中与变量 city 进行数据绑定,当下拉框选取的城市改变的时候,city 的值就变成所选城市对应的 value 值。案例刚执行的效果如图 3-26 所示,数据选择之后执行的效果如图 3-27 所示。

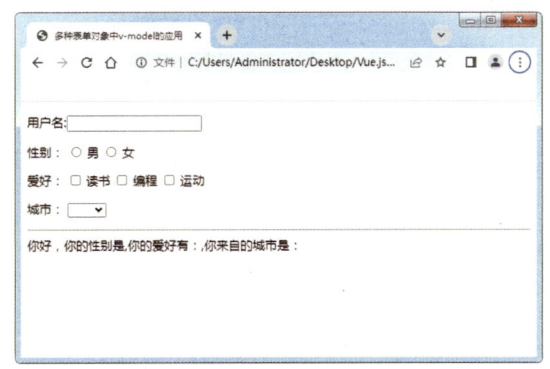

图 3-26　表单对象 v-model 的应用 1

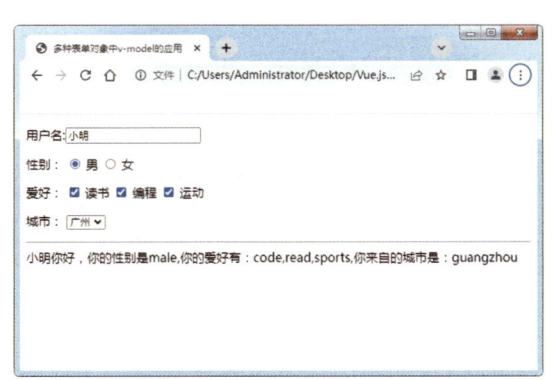

图 3-27　表单对象 v-model 的应用 2

3.6.3 制作二级级联菜单

级联菜单是指有多个层级的菜单，每个级别的一级菜单都有与之相关联的子级菜单。如图 3-28 所示，当选择一级菜单"前端基础课程"时，它的二级菜单有两个，分别为"HTML5"和"CSS3"。如图 3-29 所示，当选择一级菜单"前端进阶课程"时，它的二级菜单有两个，分别为"JavaScript"和"Vue.js"。

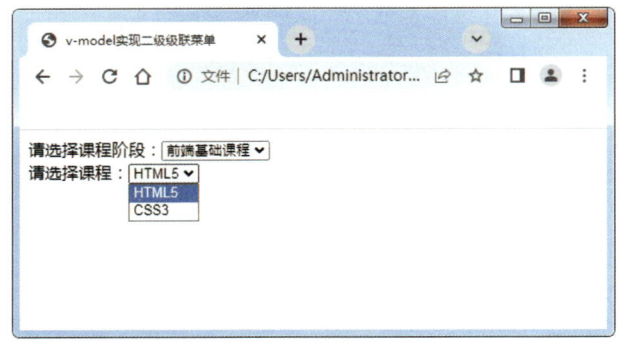

图 3-28　二级级联菜单 1

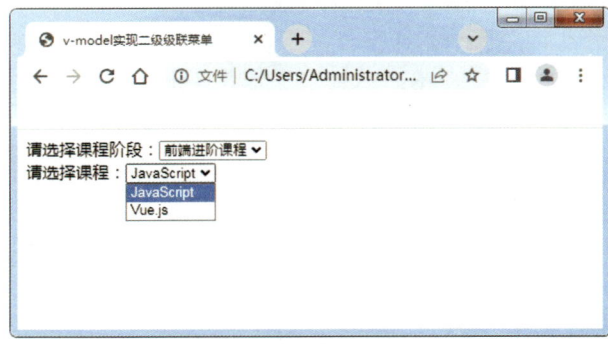

图 3-29　二级级联菜单 2

实现级联菜单对数据的设计很关键，菜单数据使用数组结构，二级菜单作为一级菜单的子数据，数据结构设计如下：

```
kecheng: [
    {
        label:"前端基础课程",
        value:"jichu",
        children: [
            {
                label: "HTML5",
                value:"html"
            },
            {
                label: "CSS3",
                value:"css"
            },
        ]
    },
    {
        label:"前端进阶课程",
        value:"jinjie",
        children: [
            {
                label: "JavaScript",
                value:"js"
            },
            {
```

```
                    label: "Vue.js",
                    value:"vue"
                    },
                ]
            },
        ]
```

二级级联菜单实现的核心是当一级目录选择好以后，二级目录要做出动态改变，这里使用了 computed 计算属性和 filter 函数来实现。filter( ) 方法用于把数组按照某个要求进行过滤，以返回值的形式将符合要求的数据 return 出来，返回值的类型也是一个数组。例如，可以通过以下代码将 1~10 中的偶数过滤出来。

```
<script>
    var  arr = [1, 2, 3, 4, 5, 6, 7, 8, 9, 10]
    var  newArr = arr.filter((item) => {//item是数组的每一项，依次进行过滤
        return  item % 2 == 0;
    })
    console.log(newArr); // 输出结果：[2, 4, 6, 8, 10]
</script>
```

案例代码如下：

```
<!DOCTYPE html>
<html lang="en">
<head>
    <meta charset="UTF-8">
    <meta http-equiv="X-UA-Compatible" content="IE=edge">
    <meta name="viewport" content="width=device-width, initial-scale=1.0">
    <title> v-model 实现二级级联菜单 </title>
    <script src="./js/vue3.js"></script>
    <script>
        window.onload = function () {
            const app = {
                data() {
                    return {
                        selected: "jichu",
                        kecheng: [
                            {
                                label: "前端基础课程",
                                value:"jichu",
                                children: [
                                    {
                                        label: "HTML5",
```

```
                value:"html"
              },
              {
                label: "CSS3",
                value:"css"
              },
            ]
          },
          {
            label:"前端进阶课程",
            value:"jinjie",
            children: [
              {
                label: "JavaScript",
                value:"js"
              },
              {
                label: "Vue.js",
                value:"vue"
              },
            ]
          },
        ]
      },
      computed: {
        selection() {
          return this.kecheng.filter((item) => {
            return item.value == this.selected;
          })[0].children;
        }
      },
    }
    Vue.createApp(app).mount('#container')
  </script>
</head>
<body>
  <div id="container">
```

```
        <span>请选择课程阶段：</span>
        <select v-model="selected">
            <option v-for="ke in kecheng" :value="ke.value">
                {{ke.label}}
            </option>
        </select>
         <br>
        <span>请选择课程：</span>
        <select>
            <option v-for="subke in selection" :value="subke.label" >
                {{subke.label}}
            </option>
        </select>
    </body>
</html>
```

> 循环出一级菜单，变量 selected 的值是下拉菜单选中项的 value 值

> selection 在 computed 中进行定义，在 option 中循环出二级菜单

一级下拉菜单与 data 中的变量 selected 进行双向数据绑定，当一级下拉菜单的值改变时，在 computed 中对变量 kecheng 进行过滤，返回 kecheng 变量的 value 值与变量 selected 相同的元素。

课后练习题

1. 与插值表达式 {{}} 的作用是等同的指令是_____。两者的区别是：在渲染的数据比较多的时候或者网络加载不好的情况下，插值表达式 {{}} 可能会先把大括号显示出来，等数据全部加载出来以后才正常显示。解决的方式是使用_____指令。

2. 用来给元素添加事件的指令是_____，该指令可以简写为_____。

3. 常用的事件修饰符有 prevent、stop、once 等。其中，_____的作用是阻止默认事件，_____的作用是阻止事件冒泡，_____的作用是只触发一次事件。

4. v-show 指令与 v-if 指令类同，区别在于_____是会从 DOM 中添加或删除元素，而_____不会从 DOM 中删除元素，而是使用 CSS 的 display 属性来控制元素的显示或隐藏。对于频繁切换的元素优先选择_____。

5. 用于循环可迭代的数据，例如数组和对象等使用_____指令。

6. 属性绑定使用_____指令。实现数据的双向绑定使用_____指令。

第 4 章

Vue 内置指令综合演练

🕐 导 言

　　伟大复兴，壮丽航程！我国在各方面都取得了举世瞩目的成就。科技是第一生产力，我国在科技领域不断涌现世界级科技成果，创新能力持续提升。科技创新快速发展，中国正向着世界科技强国的宏伟目标大步迈进。在本章中我们将运用 Vue 的各种指令完成综合案例。在焦点图效果中将会展现我国科技的几项瞩目成就。先进的科技成果是提振综合国力的国之重器，也是实现美好生活的"加速器"。

📋 学习内容

　　本章一共有 3 节，涉及的案例有：工作计划表、图书管理系统、焦点图效果。4.1 节的工作计划表使用了 v-if、v-model 等指令以及数组的增加、删除的相关方法。4.2 节的图书管理系统涉及数据的增、删、改、查等操作。4.3 节的焦点图效果使用了类的绑定以及 v-if 等指令。

📋 学习目标

1. 掌握数组的增加、删除的方法。
2. 掌握 Vue 各种内置指令的使用方法。
3. 能够分析项目需求所使用的知识点，以及利用所学知识点进行综合应用。
4. 理解并掌握类的动态绑定。

🎯 学习重点

1. 掌握 Vue 中数据增、删、改、查的方法。
2. 能够分析项目需求所使用的知识点，以及利用所学知识点进行综合应用。
3. 理解并掌握类的动态绑定。

4.1 工作计划表

4.1.1 项目描述

该项目的完成界面如图 4-1 所示。页面结构分为三个部分：计划输入部分、计划展示部分、计划信息部分。在项目中要实现计划的添加、计划的显示及计划的删除。

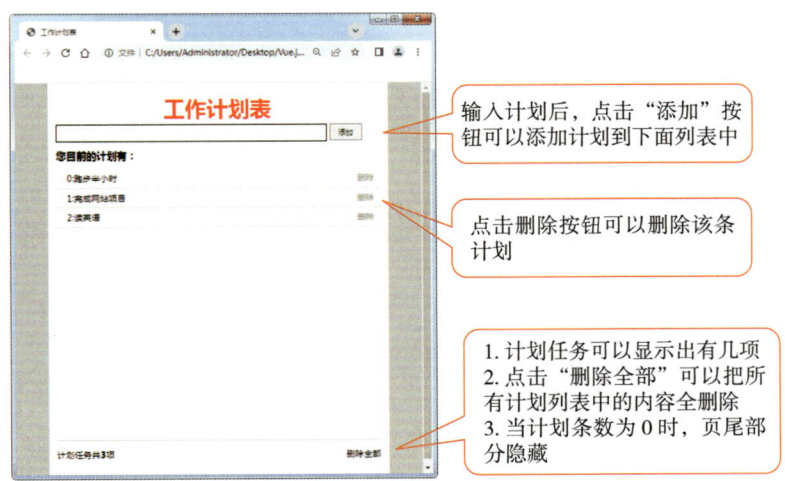

图 4-1 工作计划表完整效果

4.1.2 前置知识——数组中添加和删除数据的方法

数组中添加数据的方法有 unshift 和 push，其中 unshift 是在数组的最前面添加一条数据，push 是在数组的最后面添加一条数据。删除数据的方法有 shift 和 pop，其中 shift 是删除数组最前面的一条数据，pop 是删除数组中最后面的一条数据。splice 可以指定位置进行删除，splice 的第一个参数是要删除值的索引位置，第二个参数是要从索引值处向后删除几条数据，如果有第三个数据则是在删除值的位置上添加的新数据。示例代码如下：

```
<!DOCTYPE html>
<html lang="en">
<head>
    <meta charset="UTF-8">
    <meta http-equiv="X-UA-Compatible" content="IE=edge">
    <meta name="viewport" content="width=device-width, initial-scale=1.0">
    <title> 数组的添加删除 </title>
    <style>
        input{ margin-right: 10px;}
    </style>
    <script src="./js/vue3.js"></script>
    <script>
```

```
window.onload=function(){
  const app={
    data(){
      return {
        fruit:['苹果','西瓜','柠檬','香蕉']
      }
    },
    methods:{
      addPre(){
        this.fruit.unshift('前面添加数据：新水果')
      },
      addEnd(){
        this.fruit.push('后面添加数据：新水果')
      },
      delPre(){
        this.fruit.shift()
      },
      delEnd(){

        this.fruit.pop()
      },
      del(n){
        this.fruit.splice(n,1)
      }
    },
  }
  Vue.createApp(app).mount('#container')
}
</script>
</head>
<body>
  <div id="container">
    <p v-for="(item,index) in fruit">{{index}}------{{item}}</p>
    <input type="button" value="前面添加一条数据" @click="addPre">
    <input type="button" value="后面添加一条数据" @click="addEnd">
    <input type="button" value="前面删除一条数据" @click="delPre">
    <input type="button" value="后面删除一条数据" @click="delEnd">
    <input type="button" value="删除编号为2的数据" @click="del(2)">
  </div>
```

注释：
- n 接收 html 代码中 @click="del（2）" 里面的 2
- splice（2,1）的含义是从数组编号为2的数据开始删除，一共删除1个数据
- 这里的 2 会作为参数传递给 del（n）

```
</body>
</html>
```

代码执行效果如图 4-2 所示。点击按钮"前面添加一条数据"后的效果如图 4-3 所示。点击按钮"后面添加一条数据"后的效果如图 4-4 所示。点击按钮"前面删除一条数据"后的效果如图 4-5 所示。点击按钮"后面删除一条数据"后的效果如图 4-6 所示。点击按钮"删除编号为 2 的数据"后的效果如图 4-7 所示。

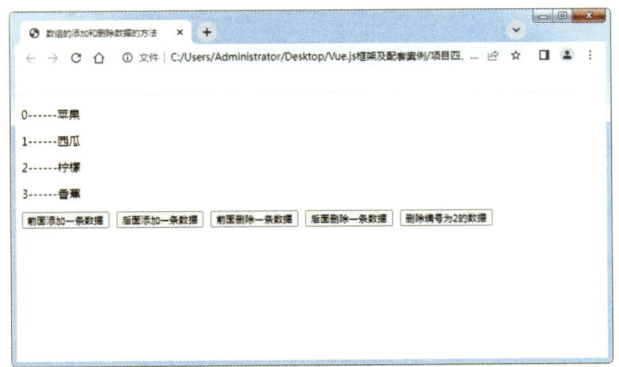

图 4-2　初始状态

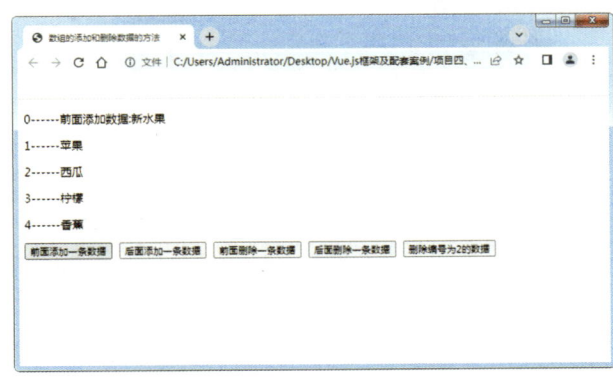

图 4-3　前面添加一条数据

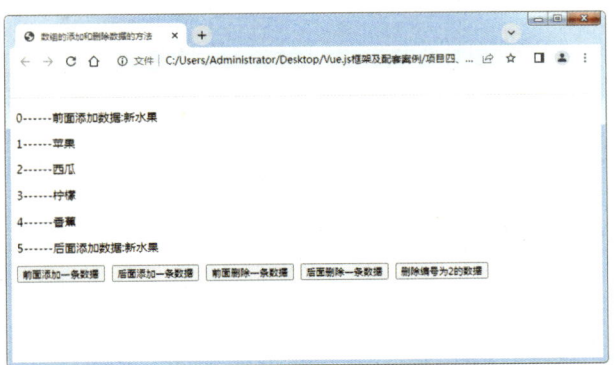

图 4-4　后面添加一条数据

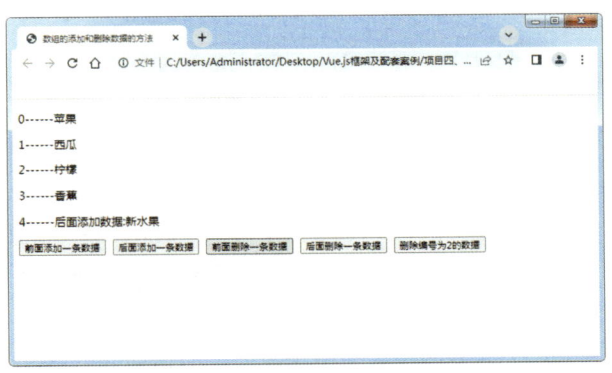

图 4-5　前面删除一条数据

图 4-6　后面删除一条数据

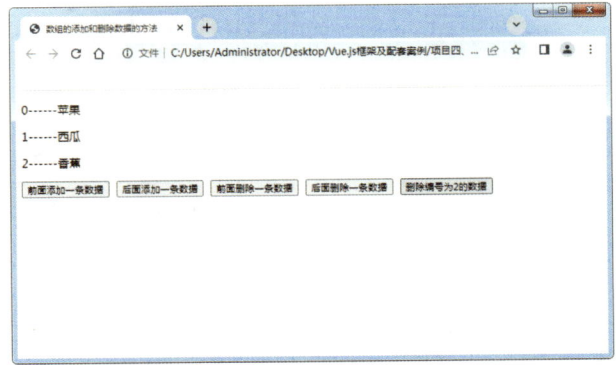

图 4-7　删除编号为 2 的数据

4.1.3　项目分析

工作计划表项目的实现思路是通过数组来存储计划列表中的内容，列表通过 v-for 循环来显示出计划任务。当点击"添加"按钮时，将文本框的计划内容通过 push 方法将新数据添加到数组中。当在每一条计划任务上点击"删除"时，通过 splice 方法删除数组中的该条数据。点击"删除全部"将数组中的所有数据清空。页尾部分通过 v-if 进行显示和隐藏。

4.1.4 项目实施

项目实施步骤 1：完成工作计划表静态布局

首先使用 HTML 和 CSS 相关知识完成该项目的界面。在静态布局中所有的数据都先使用固定的值进行占位，后面再通过数据对占位的部分进行替换。具体代码如下：

```html
<!DOCTYPE html>
<html lang="en">
<head>
    <meta charset="UTF-8">
    <meta http-equiv="X-UA-Compatible" content="IE=edge">
    <meta name="viewport" content="width=device-width, initial-scale=1.0">
    <title>工作计划表</title>
    <link rel="stylesheet" href="./style.css">
</head>
<body>
    <div id="container">
        <div id="plan">
            <h1>工作计划表</h1>
            <input type="text" class="txt" >
            <input type="button" class="btn" value="添加">
            <h3>您目前的计划有：</h3>
            <ul>
                <li>1：计划 1<span>删除</span></li>
                <li>2：计划 2<span>删除</span></li>
                <li>3：计划 3<span>删除</span></li>
            </ul>
            <div class="footer">
                <p class="left">计划任务共<strong>3</strong>项</p>
                <p class="right">删除全部</p>
            </div>
        </div>
    </div>
</body>
</html>
```

该案例的样式使用了外部链接 style.css。style.css 的代码为：

```css
*{ margin:0; padding: 0 }
body{background: #CCC;font-family: "微软雅黑"}
```

```css
#plan{ width: 600px; background: #FFF; height: 700px; margin:0 auto; padding: 15px; position: relative; }
#plan h1{ text-align: center; color: #F00; font-size: 40px; line-height: 60px }
#plan .txt{ width: 500px; height: 30px;  border:1px  #000 solid;outline: none }
#plan .btn{ width: 60px; height: 32px; margin-left: 5px; }
#plan h3{ line-height: 50px; }
#plan li{ list-style: none; padding-left: 20px; line-height: 35px; border-bottom: 1px #CCC dashed }
#plan li.on{ border:2px #C00 dashed; }
#plan li span{float: right; margin-right: 10px; color: #999; cursor: pointer;}
#plan li span:hover{ color: #000 }
#plan .footer{ position: absolute; left:15px; bottom: 20px; width: 600px;
      border-top:1px #CCC solid; line-height: 50px; }
#plan .footer .left{float: left;}
#plan .footer .right{ float: right; cursor: pointer; }
```

项目实施步骤 2：完成工作计划表添加功能

定义一个数组变量来存放工作计划，为了测试方便，先将数组里面放两个初始值，项目全部完成以后将该数组清空即可。具体代码如下：

```js
data(){
        return {
            planList:["读书","编程"],// 预置两个初始值，用于测试
            newPlan:"", // 与文本框进行双向数据绑定
        }
}
```

项目实施步骤 2：完成工作计划表添加功能

改写 HTML 结构，将 li 列表部分改写为 v-for 的形式，把 planList 数据通过循环展现出来。

点击"添加按钮"将新写在文本框里面的内容添加到 planList 数组中。同时，在文本框内输入完成后，按 Enter 键也同样可以完成添加功能。完整代码如下：

```html
<!DOCTYPE html>
<html lang="en">
<head>
    <meta charset="UTF-8">
    <meta http-equiv="X-UA-Compatible" content="IE=edge">
    <meta name="viewport" content="width=device-width, initial-scale=1.0">
    <title>工作计划表</title>
    <link rel="stylesheet" href="./css/style.css">
    <script src="./js/vue3.js"></script>
    <script>
        window.onload=function(){
            const app={
```

```
            data(){
                return {
                    planList:["读书","编程"],
                    newPlan:"",
                }
            },
            methods:{
                add (){
                    if(this.newPlan!=""){
                        this.planList.push(this.newPlan)
                        this.newPlan=""
                    }
                },
            },
        }
        Vue.createApp(app).mount('#container')
    }
    </script>
</head>
<body>
    <div id="container">
        <div id="plan">
            <h1> 工作计划表 </h1>
                <input type="text" class="txt" v-model="newPlan" @keydown.enter="add">
            <input type="button" class="btn" value=" 添加 " @click="add">
            <h3>您目前的计划有：</h3>
            <ul>
                <li v-for="(value,index) in planList">{{index}}:{{value}}
                <span>删除 </span>
                </li>
            </ul>
            <div class="footer">
                <p class="left"> 计划任务共 <strong>3</strong> 项 </p>
                <p class="right">删除全部 </p>
            </div>
        </div>
    </div>
</body>
</html>
```

> 如果文本框中的内容不为空，将文本框中的内容添加到 planList 数组，添加完毕以后将文本框清空

测试：当在文本框中输入"跑步"按 Enter 键或者点击"添加"按钮时候的效果如图 4-8 所示。

图 4-8　工作计划表添加数据

项目实施步骤 3：完成删除功能和页尾部分

项目实施步骤 3：完成删除功能和页尾部分

删除功能使用的方法是 splice，循环出来的列表中的 index 是每一个循环对象的索引值，这个索引值要作为参数传递到删除函数 del( ) 中。

```
<li v-for="(value,index) in planList">
{{index}}:{{value}}<span @click="del(index)"> 删除 </span>
</li>
```

完整代码如下：

```
<!DOCTYPE html>
<html lang="en">
<head>
    <meta charset="UTF-8">
    <meta http-equiv="X-UA-Compatible" content="IE=edge">
    <meta name="viewport" content="width=device-width, initial-scale=1.0">
    <title> 工作计划表 </title>
    <link rel="stylesheet" href="./css/style.css">
    <script src="./js/vue3.js"></script>
    <script>
        window.onload=function(){
            const app={
                data(){
                    return {
                        planList:[],         ← planList 的测试功能已经完成，把数据清空
                        newPlan:"",
                    }
                },
```

```
            methods:{
                add(){
                    if(this.newPlan!=''){
                        this.planList.push(this.newPlan)
                        this.newPlan=''
                    }
                },
                del(i){                    // i用于接收span标签中传递过来的index值
                    this.planList.splice(i,1)
                },
                delAll(){
                    this.planList=[]       // 清空planList数组，完成删除全部的功能
                },
            },
        }
        Vue.createApp(app).mount('#container')
    </script>
</head>
<body>
    <div id="container">
        <div id="plan">
            <h1>工作计划表</h1>
            <input type="text" class="txt" v-model="newPlan" @keydown.enter="add">
            <input type="button" class="btn" value="添加" @click="add">
            <h3>您目前的计划有：</h3>
            <ul>
                <li v-for="(value,index) in planList">
                    {{index}}:{{value}}<span @click="del(index)">删除</span>
                </li>
            </ul>
            <div class="footer" v-if="planList.length>0">    <!-- planList的长度大于0时footer显示 -->
                <p class="left">计划任务共<strong>{{planList.length}}</strong>项</p>
                <p class="right" @click="delAll()">删除全部</p>
            </div>
        </div>
    </div>
</body>
</html>
```

网页初始时的效果如图 4-9 所示，添加两条工作计划以后的效果如图 4-10 所示。

图 4-9 计划表初始状态

图 4-10 添加两条数据

4.2 图书管理系统

4.2.1 项目描述

图书管理系统界面使用 Bootstrap 框架完成，从功能上要完成数据的增、删、改、查。效果如图 4-11 所示。

图 4-11 图书管理系统完整效果

4.2.2 前置知识

在该项目中要使用到以下方法：forEach、findIndex、indexOf、filter，其中 filter 的用法请参看【3.6.3 制作二级级联菜单】。下面依次讲解 forEach、findIndex、indexOf 方法。

forEach可以循环数组中的每一项，对每一项中的数据进行修改，下面案例的效果是在每一位学生的ID号前面添加"2023"，具体代码如下：

```html
<!DOCTYPE html>
<html lang="en">
<head>
    <meta charset="UTF-8">
    <meta http-equiv="X-UA-Compatible" content="IE=edge">
    <meta name="viewport" content="width=device-width, initial-scale=1.0">
    <title>forEach方法</title>
    <script>
        var stu=[
            {
                name:"王小明",
                id:"01"
            },
            {
                name:"李小红",
                id:"02"
            }
        ]
        stu.forEach (item=>{          // 这里使用箭头函数，item指的是数组中的每一项
            item.id="2023"+item.id    // 在每一项中的ID值前面添加上"2023"作为新ID值
        })
        console.log(stu)
    </script>
</head>
<body>
</body>
</html>
```

案例效果如图4-12所示。

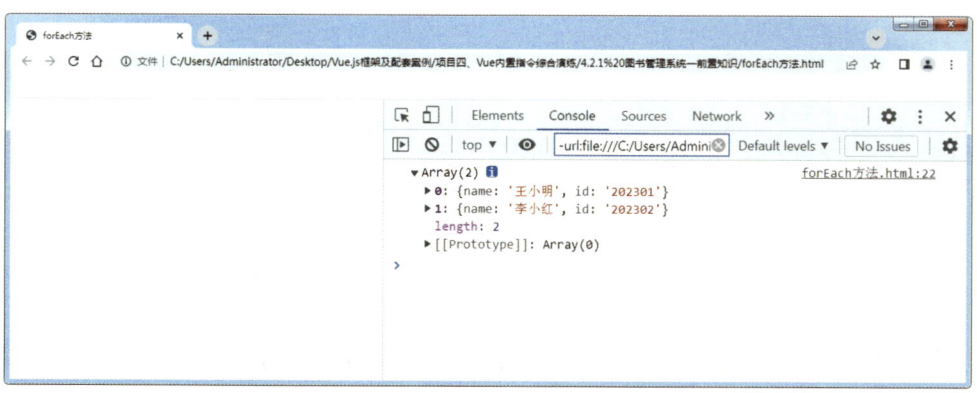

图4-12 forEach方法的案例效果

findIndex方法是通过依次遍历数组的每一项来查找符合条件的项,通过return将该项的索引值返回出去。下面案例通过查找变量stu中每一项,返回name值是"李小红"的索引值,具体代码如下:

```html
<!DOCTYPE html>
<html lang="en">
<head>
    <meta charset="UTF-8">
    <meta http-equiv="X-UA-Compatible" content="IE=edge">
    <meta name="viewport" content="width=device-width, initial-scale=1.0">
    <title>Document</title>
    <script>
        var stu=[
            {
                name:"王小明",
                id:"01"
            },
            {
                name:"李小红",
                id:"02"
            }
        ]
        var i= stu.findIndex(item=>{
          return item.name=="李小红"
        })
        console.log(i)//返回结果为1
    </script>
</head>
<body>
</body>
</html>
```

> item是数组中每一项,查找出name值是"李小红"的索引值,将索引值赋值给i

indexOf方法可以查找字符串中某个值第一次出现的位置,如果没有查找到则返回–1。下面案例在字符串"vue.js是一个JavaScript渐进式框架"中查找字符串"js"第一次出现的位置,具体代码如下:

```html
<!DOCTYPE html>
<html lang="en">
<head>
    <meta charset="UTF-8">
    <meta http-equiv="X-UA-Compatible" content="IE=edge">
    <meta name="viewport" content="width=device-width, initial-scale=1.0">
    <title>Document</title>
```

```
<script>
    var str="vue.js 是一个 JavaScript 渐进式框架"
    var i=str.indexOf("js")//返回字符串"js"在变量str中第一次出现的位置
    var j=str.indexOf("html")
    var m=str.indexOf("")
    console.log(i)//i 的值是 4
    console.log(j)//j 的值是 -1,因为从 str 中没有找到字符串"html"
    console.log(m)//m 的值为 0,默认空字符串在字符串的排序为 0
</script>
</head>
<body>

</body>
</html>
```

4.2.3 项目分析

该项目中数据的增加使用 push 方法,添加与修改共同使用同一个文本框,因此要使用 if 语句进行判断是要进行哪种操作。删除功能使用 splice,这里通过 findIndex 方法来查找到所删除对象的索引值,然后进行删除。查找功能使用 filter 和 indexOf 方法来实现。思路是使用 indexOf 方法对搜索关键词进行查找,如果值不是 -1,则证明有该关键词。

4.2.4 项目实施

项目实施步骤 1:完成图书管理系统静态布局

该项目中的静态布局使用的是 Bootstrap 框架。Bootstrap 是一个 CSS/HTML 响应式框架,网站开发者只要调用 Bootstrap 中对应的类就可以快速完成页面中的布局。Bootstrap 中各种类的具体使用方法可以参考"Bootstrap 中文网",网址为 https://www.bootcss.com/,这里不再赘述。图书管理系统静态布局的代码如下:

```
<!DOCTYPE html>
<html lang="en">
<head>
    <meta charset="UTF-8">
    <meta http-equiv="X-UA-Compatible" content="IE=edge">
    <meta name="viewport" content="width=device-width, initial-scale=1.0">
    <title>图书管理系统</title>
    <link rel="stylesheet" href="./css/bootstrap.min.css">
</head>
<body>
    <div class="container">
```

```html
<h1 class="text-center">图书管理系统</h1>
<br><br>
<form action="" class="form-inline">
    <div class="form-group">
        <label for="id">ID:</label>
        <input type="text" id="id" class="form-control">
    </div>
      
    <div class="form-group">
        <label for="name">Name:</label>
        <input type="text" id="name" class="form-control">
    </div>
      
    <input type="button" class="btn btn-info"  value="提交">
    <div class="form-group pull-right">
        <label for="sou" class="text-danger">搜索：</label>
        <input type="text" id="sou" class="form-control">
    </div>
</form>
<br><br>

<table class="table table-bordered table-hover table-striped">
    <tr>
        <th>ID</th>
        <th>Name</th>
        <th>Date</th>
        <th>操作</th>
    </tr>
    <tr>
        <td>01</td>
        <td>Javascript</td>
        <td>具体日期</td>
        <td>
            <input type="button" class="btn btn-default btn-sm"  value="修改">
              
            <input type="button" class="btn btn-default btn-sm"  value="删除">
        </td>
    </tr>
    <tr>
```

```
            <td>02</td>
            <td>Vue.js框架</td>
            <td>具体日期</td>
            <td>
                <input type="button" class="btn btn-default btn-sm" value="修改">
                  
                <input type="button" class="btn btn-default btn-sm" value="删除">
            </td>
        </tr>
    </table>
  </div>
</body>
</html>
```

图书管理系统静态界面的效果如图4-13所示。

项目实施步骤2：完成添加和修改功能

图4-13 图书管理系统静态界面

项目实施步骤2：完成添加和修改功能

为了测试方便，在data中对数组先预置两个数据，方便在HTML结构中测试v-for循环。

由于要记录的数据包括id、name、date，所以使用JSON对象来存储对应的数据。具体结构如下：

```
books:[
        {id:"001",name:"js网页特效",date:new Date().toLocaleString()},
        {id:"002",name:"vue.js框架",date:new Date().toLocaleString()},
    ]
```

toLocaleString()是将当前时间转换为字符串形式

点击"提交"按钮可以把数据添加到books数组中，但是因为点击"修改"按钮要实现回填到文本框的功能，其效果如图4-14所示，所以设置变量flag用于判断是新内容的添加还是旧内容的修改。设定flag的值初始为false，当点击"修改"按钮时flag的值变为true。

第4章　Vue内置指令综合演练

图4-14　点击修改按钮内容回填到文本框

完成添加和修改功能的代码如下：

```
<!DOCTYPE html>
<html lang="en">

<head>
    <meta charset="UTF-8">
    <meta http-equiv="X-UA-Compatible" content="IE=edge">
    <meta name="viewport" content="width=device-width, initial-scale=1.0">
    <title>图书管理系统添加和修改功能</title>
    <link rel="stylesheet" href="./css/bootstrap.min.css">
    <script src="./js/vue3.js"></script>
    <script>
        window.onload = function () {
            const app = {
                data() {
                    return {
                        books: [
                            { id: "001",
                              name: "js网页特效",
                              date: new Date().toLocaleString()
                            },
                            { id: "002",
                              name: "HTML+CSS",
                              date: new Date().toLocaleString()
                            },
                        ],
                        id: "",
```

77

```js
            name: '',                    // id 和 name 分别与下面的文本框双
            flag: false,                 // 向数据绑定，flag 用于标记是添加
        }                                // 还是修改
    },
    methods: {
        tijiao() {
            if (this.flag == false) {    // 添加新内容
                if (this.id != "" && this.name != "") {
                    xin = {
                        id: this.id,
                        name: this.name,
                        date: new Date().toLocaleString()
                    }
                    this.books.push(xin)
                    this.id = ''// 清空填入的 id 值
                    this.name = ''// 清空填入的 name 值
                }
            }
            else {                       // 修改旧内容
                this.books.forEach(item => {// 循环 books 数组
                    if (item.id == this.id) {// 找到要修改的那一项
                        item.name = this.name// 修改 name 值
                        this.flag = false// 将 flag 的值重新改为 false
                        this.id = ''
                        this.name = ''
                    }
                })
            }
        },
        fill(id) {                       // 将内容回填到文本框中
            selected = this.books.filter((item) => {
                return item.id == id
                // 通过查找 id 找到要修改的项存在 selected 数组中
            })
            this.flag = true
            this.id = selected[0].id// 将内容填到 id 文本框
            this.name = selected[0].name// 将内容填到 name 文本框
        },
```

```
                },
            }
                Vue.createApp(app).mount('.container')
        }
    </script>
</head>

<body>
    <div class="container">
        <h1 class="text-center">图书管理系统</h1>
        <br><br>
        <form action="" class="form-inline">
            <div class="form-group">
                <label for="id">ID:</label>
                <input type="text" id="id" class="form-control"
                v-model="id" :disabled="flag" >
            </div>
              
            <div class="form-group">
                <label for="name">Name:</label>
                <input type="text" id="name" class="form-control" v-model="name">
            </div>
              
            <input type="button" class="btn btn-info" value="提交" @click="tijiao()">
            <div class="form-group pull-right">
                <label for="sou" class="text-danger">搜索：</label>
                <input type="text" id="sou" class="form-control">
            </div>
        </form>
        <br><br>
        <table class="table table-bordered table-hover table-striped">
            <tr>
                <th>ID</th>
                <th>Name</th>
                <th>Date</th>
                <th>操作</th>
            </tr>
            <tr v-for="value in books">
                <td>{{value.id}}</td>
```

> 进行修改的时候，id 文本框不可用，id 值是不能被修改的

```html
                    <td>{{value.name}}</td>
                    <td>{{value.date}}</td>
                    <td>
                        <input type="button" class="btn btn-default btn-sm"
                         @click="fill(value.id)" value="修改">
                          
                        <input type="button" class="btn btn-default btn-sm"  value="删除">
                    </td>
                </tr>
        </table>
    </div>
</body>
</html>
```

点击修改，执行 fill 函数

项目实施步骤 3：完成删除和搜索功能

删除操作通过 findeIndex 方法获取要删除项的索引值，然后通过 splice 将该项删除。搜索功能的实现要定义一个变量与搜索框双向数据绑定，使用 filter 将符合要求的值以数组的形式返回。具体代码如下：

```html
<!DOCTYPE html>
<html lang="en">
<head>
    <meta charset="UTF-8">
    <meta http-equiv="X-UA-Compatible" content="IE=edge">
    <meta name="viewport" content="width=device-width, initial-scale=1.0">
    <title>图书管理系统添加和修改功能</title>
    <link rel="stylesheet" href="./css/bootstrap.min.css">
    <script src="./js/vue3.js"></script>
    <script>
        window.onload = function () {
            const app = {
                data() {
                    return {
                        books: [],
                        id: '',
                        name: '',
                        flag: false,
                        sear:''
                    }
                },
                methods: {
                    tijiao() {
```

项目实施步骤 3：完成删除和搜索功能

将测试值删除

与搜索框双向数据绑定

```
                    if (this.flag == false) {
                        if (this.id != "" && this.name != "") {
                            xin = { id: this.id,
                                name: this.name,
                                date: new Date().toLocaleString()
                            }
                            this.books.push(xin)
                            this.id = ''
                            this.name = ''
                        }
                    }
                    else {
                        this.books.forEach(item => {
                            if (item.id == this.id) {
                                item.name = this.name
                                this.flag = false
                                this.id = ''
                                this.name = ''
                            }
                        })
                    }
                },
                fill(id) {
                    selected = this.books.filter((item) => {
                        return item.id == id
                    })
                    this.flag = true
                    this.name = selected[0].name
                    this.id = selected[0].id
                },
                del(id){
                    const i=this.books.findIndex(item=>{
                        return item.id==id
                    })
                    this.books.splice(i,1)
                },
                search(){
                    const result=this.books.filter(item=>{
                        return  item.name.indexOf(this.sear)!=-1
                    })
```

> 循环 books 数组查看每一项的 id，找到与传入的参数 id 相同的那一项，将该项的索引值通过返回值的形式赋值给 i

> 在每一项的 name 中查找搜索框的内容出现的位置，如果不是 −1，证明包含搜索框的内容

```
                    return result
                }
            },
        }
        Vue.createApp(app).mount('.container')
    }
    </script>
</head>


<body>
    <div class="container">
        <h1 class="text-center">图书管理系统</h1>
        <br><br>
        <form action="" class="form-inline">
            <div class="form-group">
                <label for="id">ID:</label>
                <input type="text" id="id" class="form-control" v-model="id" :disabled="flag">
            </div>
              
            <div class="form-group">
                <label for="name">Name:</label>
                <input type="text" id="name" class="form-control" v-model="name">
            </div>
              
            <input type="button" class="btn btn-info" value="提交" @click="tijiao()">
            <div class="form-group pull-right">
                <label for="sou" class="text-danger">搜索：</label>
                <input type="text" id="sou" class="form-control" v-model="sear">
            </div>
        </form>
        <br><br>
        <table class="table table-bordered table-hover table-striped">
            <tr>
                <th>ID</th>
                <th>Name</th>
                <th>Date</th>
                <th>操作</th>
            </tr>
            <tr v-for="value in search()">
                <td>{{value.id}}</td>
```

> search 函数的返回值是 result

> 循环的值是 search 函数的返回值，即 result 的值

```
                    <td>{{value.name}}</td>
                    <td>{{value.date}}</td>
                    <td>
                        <input type="button" class="btn btn-default btn-sm"
                        @click="fill(value.id)" value="修改">
                          
                        <input type="button" class="btn btn-default btn-sm" value="删除"
                        @click="del(value.id)">
                    </td>
                </tr>
            </table>
        </div>
    </body>
</html>
```

在页面中添加新数据，效果如图 4-15 所示。在搜索框中输入"js"，对数据进行搜索，效果如图 4-16 所示。

图 4-15　添加 3 条数据

图 4-16　搜索"js"关键词

4.3 焦点图效果

4.3.1 项目描述

焦点图是前端设计中的一个重要组成部分，该项目效果如图 4-17 所示。焦点图效果中有三张图片进行切换，图片以及标题从 data 的数据中获取。焦点图下面的小图标的数量与图片的数量相对应，切换图片时，对应的小图标也同时变色，点击小图标可以跳转到对应的图片。

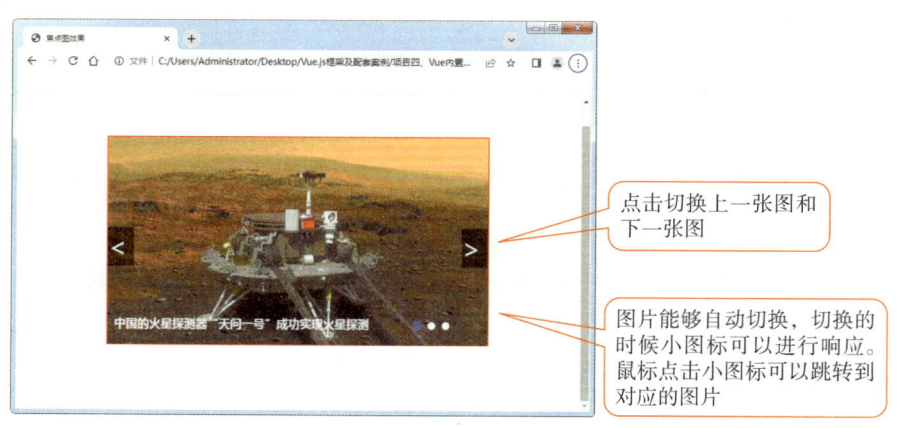

图 4-17 焦点图效果

4.3.2 前置知识

类的绑定语法结构是 :class="{ 类名 : 布尔值 }"，如果布尔值为 true，则该类生效，如果布尔值为 false，则该类不生效。下面案例的效果是点击 li 以后，这个 li 添加类 .bd。具体代码如下：

```
<!DOCTYPE html>
<html lang="en">
<head>
  <meta charset="UTF-8">
  <meta http-equiv="X-UA-Compatible" content="IE=edge">
  <meta name="viewport" content="width=device-width, initial-scale=1.0">
  <title>点击 li 添加类</title>
  <style>
    .bd {
      border: 3px red solid
    }
  </style>
  <script src="./js/vue3.js"></script>
  <script>
    window.onload = function () {
      const app = {
```

```
        data() {
            return {
                s: ['苹果','香蕉','橘子','西瓜'],
                i: 0
            }
        },
        methods: {
            change (t) {
                this.i = t        将所点击li的索引值赋值给i
            }
        }
        Vue.createApp(app).mount('#container')
    }
    </script>
</head>

<body>
    <div id="container">
        <ul>
            <li v-for="(value,index) in s" :class="{bd:index==i}" @click="change(index)">
                {{value}}----------{{index}}
            </li>
        </ul>
        <p>i 当前的值是：{{i}}</p>            直观监测 i 的值
    </div>
</body>
</html>
```

将循环出来的 li 的索引值作为 change 函数的参数

代码执行后的效果如图 4-18 所示。

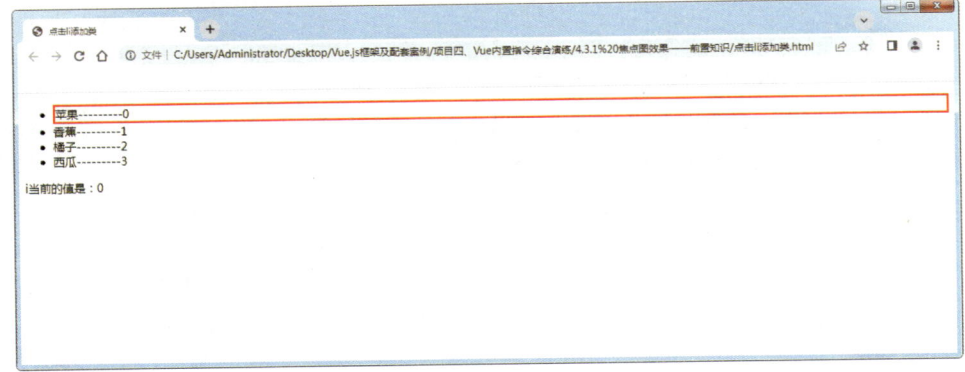

图 4-18　点击 li 添加类

该案例要深入理解类是如何进行绑定的。当点击"香蕉"时，这个 li 的索引值为 1，同时索引值通过 change 函数将 i 的值也修改为 1，执行 :class="{bd:index==i}" 指令时，index==i 的结果为 true，所以类 .bd 在香蕉这个 li 上生效，而在其他的 li 上，如以"西瓜"为例，"西瓜"的索引值 index 为 3，因此 index==i 的结果为 false，类 .bd 在这个 li 上就不会生效。

4.3.3 项目分析

设置一个变量 i 来记录当前显示图片的索引值，切换图片时改变变量 i 的值，通过 v-for 指令来动态生成小图标，通过类的绑定来实行图片切换时对应小图标状态的切换。使用生命周期 created 来实行图片的自动切换。

4.3.4 项目实施

项目实施步骤 1：完成焦点图效果静态布局

在静态布局中要对图片、图片标题以及小图标进行结构以及样式的设定，具体代码如下：

```html
<!DOCTYPE html>
<html lang="en">
<head>
    <meta charset="UTF-8">
    <meta http-equiv="X-UA-Compatible" content="IE=edge">
    <meta name="viewport" content="width=device-width, initial-scale=1.0">
    <link rel="stylesheet" href="./css/style.css">
    <title>焦点图效果</title>
</head>
<body>
    <div class="cont">
        <div class="focus">
            <div class="pic">
                <div class="item">
                    <img src="./images/pic1.jpg">
                    <h3>中国的火星探测器"天问一号"成功实现火星探测</h3>
                </div>
            </div>
            <div class="left">&lt;</div>
            <div class="right">&gt;</div>
            <div class="list">
                <ul>
                    <li class="on"></li>
                    <li></li>
                    <li></li>
                </ul>
```

```
            </div>
        </div>
    </div>
</body>
</html>
```

style.css 的代码为：

```
*{ margin: 0; padding: 0;}
.focus{ width: 590px; height: 320px; margin: 100px auto; border:2px red solid; position: relative;}
.pic{ width: 590px; height: 320px; position:absolute}
.focus .pic img{ width: 590px; height: 320px; }
.focus .pic h3{ position: absolute; left: 10px; bottom: 20px; color: #FFF; font-size: 18px;}
.focus .left,.focus .right{ width: 40px; height: 60px; background-color: rgba(0,0,0,0.5); color: #FFF; position: absolute; top:140px; font-size: 40px; line-height: 60px;  cursor: pointer;}
.focus .right{ right: 0;}
.focus .list{ width: 100px; height: 20px; position: absolute; right: 20px; bottom: 20px; }
.focus .list li{ width: 12px; height: 12px; background-color: rgb(255, 255, 255); border-radius: 50%; float: left; margin: 5px; list-style: none; cursor: pointer;}
.focus .list li.on{ background-color: rgb(43, 107, 226); transform: scale(1.3,1.3);}
```

代码执行后的效果如图 4-19 所示。

图 4-19　焦点图静态效果

项目实施步骤 2：将数据渲染到 HTML 结构中

由于数据中要存放图片的路径和图片的标题，因此使用 JSON 对象来存储。定义变量 i 的初始值为 0，用来记录当前显示图片的索引值。具体代码如下：

```html
<!DOCTYPE html>
<html lang="en">
<head>
    <meta charset="UTF-8">
    <meta http-equiv="X-UA-Compatible" content="IE=edge">
    <meta name="viewport" content="width=device-width, initial-scale=1.0">
    <link rel="stylesheet" href="./css/style.css">
    <title>焦点图效果</title>
    <script src="./js/vue3.js"></script>
    <script>
        window.onload = function () {
            const app = {
                data() {
                    return {
                        focusPic:[
                            {
                                imgSrc:"./images/pic1.jpg",
                                title:"中国的火星探测器"天问一号"成功实现火星探测"
                            },
                            {
                                imgSrc:"./images/pic2.jpg",
                                title:"中国的天文观测设备取得丰硕成果"
                            },
                            {
                                imgSrc:"./images/pic3.jpg",
                                title:"国产C919大型客机订单破千"
                            },
                        ],
                        i:0
                    }
                },
            }
            Vue.createApp(app).mount('.cont')
        }
    </script>
</head>
<body>
    <div class="cont">
        <div class="focus">
```

```html
            <div class="pic" v-for="(value,index) in focusPic">
                <div class="item" v-if="i==index">
                    <img  :src="value.imgSrc">
                    <h3>{{value.title}}</h3>
                </div>
            </div>
            <div class="left">&lt;</div>
            <div class="right">&gt;</div>
            <div class="list">
                <ul>
                    <li v-for="(value,index) in focusPic" :class="{on:index==i}"></li>
                </ul>
            </div>
        </div>
    </div>
</body>
</html>
```

> 循环出三组图片，只有 index 值与 i 值相同的那组图片才显示

> 通过循环渲染出三个小图标，点击 li 会添加上 on 类

项目实施步骤 3：完成图片的切换和最终效果

点击左箭头，切换到前一张图片，点击右箭头，切换到后一张图片。在图片切换的时候，变量 i 的值会变，所以对应的小图标就会添加上 on 类。当点击小图标时，改变 i 的值，对应的图片也得到切换。具体代码如下：

```html
<!DOCTYPE html>
<html lang="en">
<head>
    <meta charset="UTF-8">
    <meta http-equiv="X-UA-Compatible" content="IE=edge">
    <meta name="viewport" content="width=device-width, initial-scale=1.0">
    <link rel="stylesheet" href="./css/style.css">
    <title>焦点图效果</title>
    <script src="./js/vue3.js"></script>
    <script>
        window.onload = function () {
            const app = {
                data() {
                    return {
                        focusPic:[
                            {
                                imgSrc:"./images/pic1.jpg",
                                title:"中国的火星探测器"天问一号"成功实现火星探测",
```

```
                },
                {
                    imgSrc:"./images/pic2.jpg",
                    title:"中国的天文观测设备取得丰硕成果"
                },
                {
                    imgSrc:"./images/pic3.jpg",
                    title:"国产C919大型客机订单破千"
                },
            ],
            i:0
        }
    },
    methods:{
        prevPic(){
                if(this.i==0){
                    this.i=this.focusPic.length-1
                }
                else{
                    this.i--
                }
        },
        nextPic(){
            if(this.i==this.focusPic.length-1){
                this.i=0
            }
            else{
                this.i++
            }
        },
        startPic(){
            this.timer = setInterval(()=>{
                this.nextPic()
            },2000)
        },
        stopPic(){
            clearInterval(this.timer)
        }
    },
```

```
                created(){
                    this.startPic()
                }
            }
            Vue.createApp(app).mount('.cont')
        }
    </script>
</head>
<body>
    <div class="cont">
        <div class="focus"  @mouseover="stopPic" @mouseout="startPic">
            <div class="pic" v-for="(value,index) in focusPic">
                <div class="item" v-if="i==index">
                    <img  :src="value.imgSrc">
                    <h3>{{value.title}}</h3>
                </div>
            </div>
            <div class="left" @click="prevPic()">&lt;</div>
            <div class="right" @click="nextPic()">&gt;</div>
            <div class="list">
                <ul>
                    <li v-for="(value,index) in focusPic" :class="{on:index==i}"
                    @click="i=index">
                    </li>
                </ul>
            </div>
        </div>
    </div>
</body>
</html>
```

> 鼠标经过 .pic 时，定时器关闭；离开 .pic 时，定时器开始

> 点击小图标改变 i 的值，i 值改变显示的图片就相应改变

焦点图效果完成后，图片可以自动切换，同时点击小图标也可以切换到对应的图片，其效果如图 4-20 和图 4-21 所示。

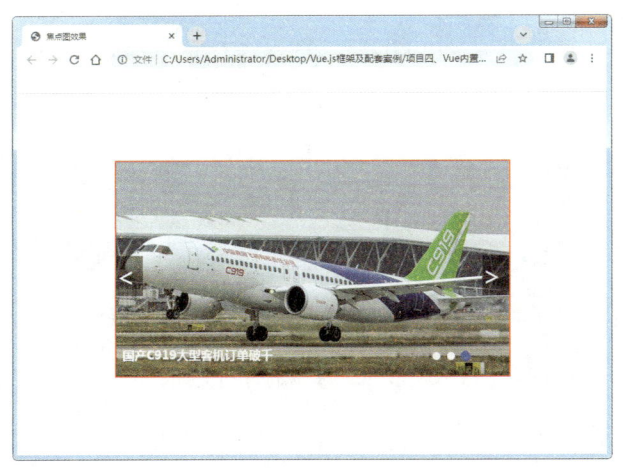

图 4-20　点击进行图片切换 1

图 4-21　点击进行图片切换 2

课后练习题

1. 数组中添加数据的方法有_____和_____，其中_____是在数组的最前面添加一条数据，_____是在数组的最后面添加一条数据。

2. 删除数据的方法有_____和_____，其中_____是删除数组最前面的一条数据，_____是删除数组中最后面的一条数据。

3. 可以指定位置进行删除的方法是_____，该方法的第一个参数是要删除值的索引位置，第二个参数是要从索引值处向后删除几条数据。

4. 在 forEach、findIndex、indexOf 几个方法中，_____可以循环数组中的每一项。_____是通过依次遍历数组的每一项来查找符合条件的项，通过 return 将值该项的索引值返回出去。_____可以查找字符串中某个值第一次出现的位置，如果没有查找到则返回 –1。

第 5 章

Vue 过渡与动画

导言

我国将提升全民族身体素养作为重要战略目标。习近平总书记对广大青少年提出了"文明其精神，野蛮其体魄"的殷切期望。体育锻炼能够增强体质、健全人格、锻炼意志，实现身心的全面发展。在本章中我们将学习 Vue 的过渡与动画，让页面元素可以动起来。

学习内容

本章主要讲解的内容是制作 Vue 的过渡与动画效果，一共有 3 节。5.1 节讲解的是 Vue 过渡效果的制作，过渡效果的核心是 CSS 中六个样式的设置。5.2 节讲解的 Vue 动画，Vue 动画与 Vue 过渡既有联系又有区别，可以对比学习。5.3 节讲解的是 Vue 过渡与动画的综合实例，分别是给工作计划表添加过渡效果和给焦点图添加动画效果。

学习目标

1. 理解 Vue 过渡的六个样式使用的情境。
2. 掌握 Vue 过渡效果的制作方法。
3. 掌握 Vue 动画效果的制作方法。
4. 掌握结合第三方库来制作 Vue 动画效果的方法。
5. 灵活应用 Vue 过渡与动画制作综合实例。

学习重点

1. 掌握 Vue 过渡中六个样式的使用。
2. 掌握 Vue 动画的制作方法。
3. 能够分析项目需求，灵活应用 Vue 的过渡与动画。

5.1 Vue 过渡

Vue 单个对象的过渡

5.1.1 Vue 单个对象的过渡

Vue 过渡需要将对象包含在一对 transition 标签中,这实际上是 Vue 提供的 transition 封装组件。

一个对象从隐藏的状态切换到显示的状态,可以理解为"从无到有"的一个过程,这个"从无到有"过程分为三个阶段:开始("无"的状态)、结束("有"的状态)和中间过渡。"开始"对应的类是 .v-enter-from,"结束"对应的类是 .v-enter-to,"中间过渡"对应的类是 .v-enter-active,如图 5-1 中的 Enter 部分。

一个对象从显示的状态切换到隐藏的状态,可以理解为"从有到无"的一个过程,这个"从有到无"过程分为三个阶段:开始("有"的状态)、结束("无"的状态)和中间过渡。"开始"对应的类是 .v-leave-from,"结束"对应的类是 .v-leave-to,"中间过渡"对应的类是 .v-leave-active,如图 5-1 中的 Leave 部分。

注意:Vue 3.0 版本使用的是 .v-enter-from 和 .v-leave-from,而在 Vue 2.0 版本对应的类分别是 .v-enter 和 .v-leave。要根据 Vue 的版本选择不同的类。

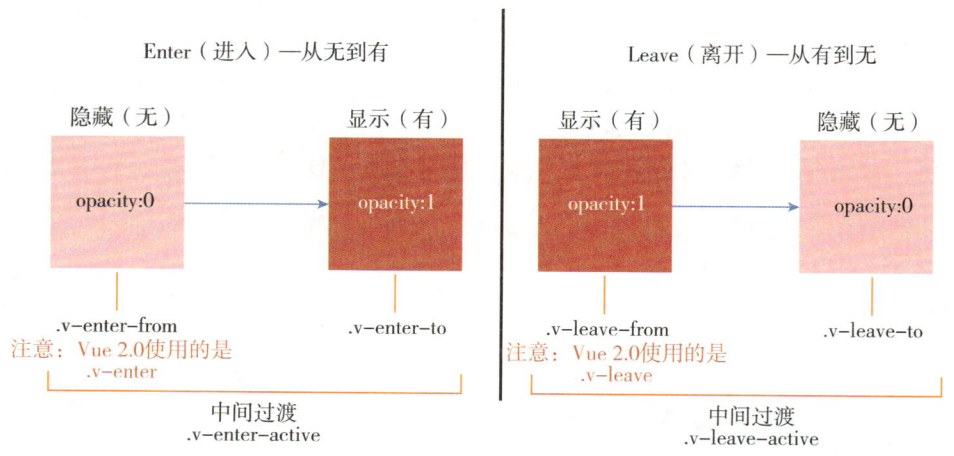

图 5-1 Vue 过渡

下面案例通过点击"切换"按钮来实现 .box 的隐藏或显示。页面刚加载出来的时候,div 是显示状态,点击"切换"按钮,div 从显示到隐藏,过渡的方式是向右移动。当 div 是隐藏状态时,点击"切换"按钮,div 从隐藏到显示,过渡的方式是通过透明度的改变来实现过渡。具体代码如下:

```
<!DOCTYPE html>
<html lang="en">
<head>
    <meta charset="UTF-8">
    <title>Vue 单个对象过渡</title>
    <style>
        .box{ width: 500px; height: 300px; background-color: green; color: #FFFFFF;
        text-align: center; line-height: 100px;}
```

```
            .v-enter-from{ opacity: 0;}
            .v-enter-to{ opacity: 1; }
            .v-enter-active{ transition: all 1s;}

            .v-leave-from{ transform: translateX(0px);}
            .v-leave-to {transform: translateX(300px)}
            .v-leave-active {transition: all 1s;}
        </style>
        <script src="js/vue3.js"></script>
        <script>
            window.onload = function () {
                const app = {
                    data() {
                        return {
                            flag: true,
                        }
                    },
                    methods: {
                        toggle () {
                            this.flag = !this.flag
                        },
                    }
                }
                Vue.createApp(app).mount('#container')
            }
        </script>
    </head>
    <body>
        <div id="container">
            <input type="button" value="切换" @click="toggle">
            <transition>
                <div v-if="flag" class="box">Vue 过渡 </div>
            </transition>
        </div>
    </body>
</html>
```

注释：
- 从隐藏到显示使用透明度进行过渡
- 从显示到隐藏使用水平平移进行过渡
- 使用 transition 包含过渡对象

从显示到隐藏使用水平平移进行过渡，隐藏过程的效果如图 5-2 所示。从隐藏到显示使用透明度进行过渡，显示过程的效果如图 5-3 所示。

图 5-2 隐藏过程的效果

图 5-3 显示过程的效果

5.1.2 对样式进行简化

从显示到隐藏的初始状态（.v-enter-from）和从隐藏到显示的最终状态（.v-leave-to），一般情况下都应该是对象默认初始的状态，所以这两个样式可以省略。

.v-enter-active 和 .v-leave-active 两个类是过渡状态的类，代码一般相同，可以合并在一起进行定义。其他代码不变，样式部分可以简化为：

```
.box{ width: 500px; height: 300px; background-color: green; color: #FFFFFF; text-align: center; line-height: 100px;}
.v-enter-from{ opacity: 0;}
.v-leave-to {transform: translateX(300px)}
.v-enter-active,.v-leave-active {transition: all 1s;}
```

5.1.3 设置多种过渡动画效果

如果页面中有多个对象需要添加过渡效果，但是每一个对象的过渡效果不同，则需要给 transition 添加一个 name 属性，CSS 中对应的过渡类也要改成这个名字作为前缀。具体代码如下：

```
<!DOCTYPE html>
<html lang="en">
<head>
    <meta charset="UTF-8">
    <meta http-equiv="X-UA-Compatible" content="IE=edge">
    <meta name="viewport" content="width=device-width, initial-scale=1.0">
    <title>设置多种过渡动画效果</title>
    <style>
        .box1{ width: 400px; height: 200px; background-color: burlywood; text-align: center; line-height: 100px; }
        .box2{ width: 400px; height: 200px; background-color:cadetblue; text-align: center; line-height: 100px;}
        .box1-enter-from,.box1-leave-to{opacity: 0; }
```

添加 box1 前缀

```
            .box1-enter-active,.box1-leave-active{ transition: all 0.5s;}

            .box2-enter-from,.box2-leave-to{transform: translateX(300px); opacity: 0;}
            .box2-enter-active,.box2-leave-active{ transition: all 0.5s;}
```

> 添加 box2 前缀

```
        </style>
        <script src="./js/vue3.js"></script>
        <script>
            window.onload=function(){
                const app = {
                    data() {
                        return {
                            flag1:true,
                            flag2:true
                        }
                    },
                }
                Vue.createApp(app).mount('.cont')
            }
        </script>
    </head>
    <body>
        <div class="cont">
            <input type="button" value="切换 box1" @click="flag1=!flag1">
            <transition name="box1">
                <div class="box1" v-if="flag1">box1</div>
            </transition>
            <hr>
            <input type="button" value="切换 box2"  @click="flag2=!flag2">
            <transition name="box2">
                <div class="box2" v-if="flag2">box2</div>
            </transition>
        </div>
    </body>
</html>
```

> 添加 name 属性，与样式前缀 box1 对应

> 添加 name 属性，与样式前缀 box2 对应

页面刚加载的初始状态如图 5-4 所示。box1 使用透明度的改变进行过渡，其效果如图 5-5 所示。box2 使用水平平移和透明度的改变进行过渡，其效果如图 5-6 所示。

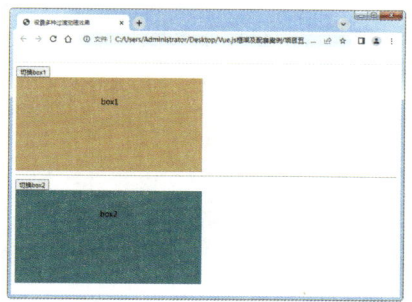
图 5-4 Vue 过渡的初始状态

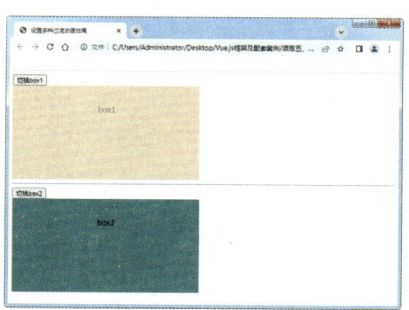
图 5-5 box1 过渡

图 5-6 box2 过渡

Vue 自定义动画

5.2 Vue 动画

Vue 动画效果与过渡效果的设置相似，区别是动画效果要使用 @keyframes 进行设置，不使用 .v-enter-from、.v-enter-to、.v-leave-from、.v-leave-to 四个类，将动画在 .v-enter-active 和 .v-leave-active 中进行调用。

5.2.1 Vue 自定义动画

下面案例点击"动画切换"按钮可以让 div 实现大小改变的动画。离开的动画效果是 div 先变大成 1.5 倍然后消失，进入的动画是离开动画的反方向播放效果，其代码如下：

```
<!DOCTYPE html>
<html lang="en">
<head>
    <meta charset="UTF-8">
    <title>Vue 单个对象动画</title>
    <style>
    .box {
    width: 200px;height: 100px;background-color: green;color: #FFFFFF;text-align: center;
    line-height: 100px;margin: 50px auto;}

    .v-enter-active {animation: bounce 0.5s;}
    .v-leave-active {animation: bounce 0.5s reverse;}

    @keyframes bounce {
        0% {
            transform: scale(0);
        }
```

> 进入和离开使用动画 bounce 效果，reverse 是反方向播放

> 使用 CSS 3.0 中的 @keyframes 进行动画 bounce 的设置

```
            50% {
                transform: scale(1.5);
            }
            100% {
                transform: scale(1);
            }
        }
    </style>
    <script src="js/vue3.js"></script>
    <script>
        window.onload = function () {
            const app = {
                data() {
                    return {
                        flag: true,
                    }
                },
                methods: {
                    toggle() {
                        this.flag = !this.flag
                    },
                }
            }
            Vue.createApp(app).mount('#container')
        }
    </script>
</head>
<body>
    <div id="container">
        <input type="button" value="动画切换" @click="toggle">
        <transition>
            <div v-if="flag" class="box">Vue 动画 </div>
        </transition>
    </div>
</body>
</html>
```

页面刚加载的初始状态如图 5-7 所示。点击"动画切换"按钮，div 先变大到 1.5 倍，其效果如图 5-8 所示，然后变小直到消失，其效果如图 5-9 所示。

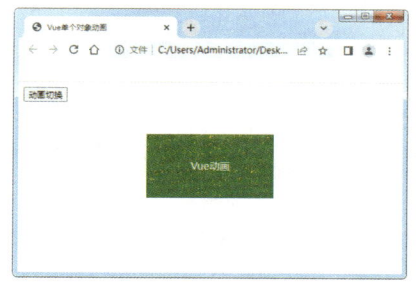 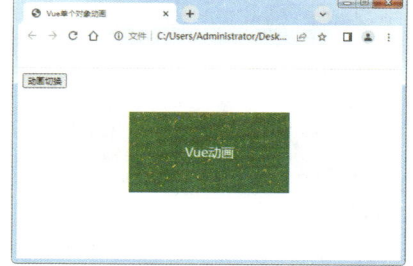 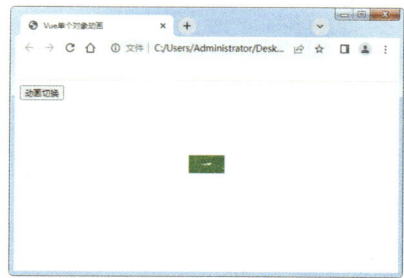

图 5-7　Vue 动画的初始状态　　　图 5-8　div 变大　　　图 5-9　div 变小

5.2.2　使用 Animate.css 动画库

Animate.css 是一个著名的动画库，里面有丰富的动画效果。Animate.css 的网址为 https://animate.style/，其网站界面如图 5-10 所示，点击右侧的类名称可以预览动画效果。

图 5-10　Animate.css 官网

使用 Animate.css 动画库时首先要引入 CSS 链接，这里使用 cdn 进行链接，然后使用对应的类就可以给对象添加上动画效果，下面的案例给 h1 标签添加了 swing 效果，其代码如下：

```
<html lang="en">
<head>
    <meta charset="UTF-8">
    <meta http-equiv="X-UA-Compatible" content="IE=edge">
    <meta name="viewport" content="width=device-width, initial-scale=1.0">
    <title>Document</title>
    <link rel="stylesheet"
    href="https://cdnjs.cloudflare.com/ajax/libs/animate.css/4.1.1/animate.min.css" />
</head>
<body>
    <h1 class="animate_animated animate_swing">animate.css</h1>
</body>
</html>
```

使用 cdn 链接 animate.min.css

必须先加上基础类 animate.min.css

具体的动画效果类 animate.min.css

在使用类的时候 animate_animated 是基础类，必须要有这个类，后面才能使用具体效果的动画类，例如 swing 效果使用的类是 animate_swing，tada 效果使用的类是 animate_tada 这里 h1 使用了 swing 类，其效果如图 5-11 所示。

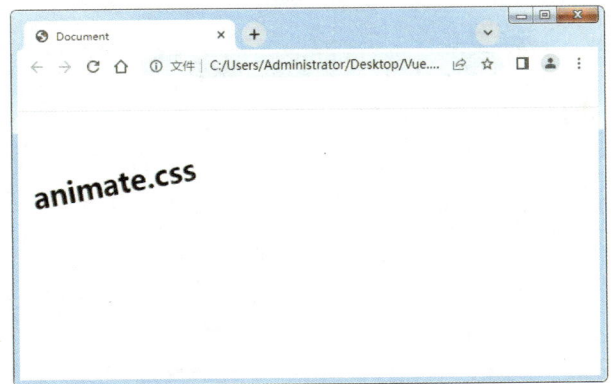

图 5-11　h1 使用 swing 动画

5.2.3　Vue 结合 Animate.css 实现动画效果

Vue 中可以给 transition 添加属性 enter-active-class 和 leave-active-class。enter-active-class 的值是进入动画使用的类，leave-active-class 的值是离开动画使用的类。下面案例中 div 进入的时候使用的效果是旋转进入，对应的类是 animate_rotateIn，离开的时候使用的效果是旋转离开，对应的类是 animate_rotateOut。其代码如下：

```
<!DOCTYPE html>
<html lang="en">
<head>
    <meta charset="UTF-8">
    <title>Vue 单个对象动画</title>
    <link rel="stylesheet" href="https://cdnjs.cloudflare.com/ajax/libs/animate.css/4.1.1/animate.min.css" />
    <style>
      .box {
        width: 300px;height: 100px;background-color: green;color: #FFFFFF;text-align: center;line-height: 100px;margin: 50px auto;
      }
    </style>
    <script src="js/vue3.js"></script>
    <script>
        window.onload = function () {
            const app = {
                data() {
                    return {
                        flag: true,
                    }
```

```
                },
                methods: {
                    toggle() {
                        this.flag = !this.flag
                    },
                }
            }
            Vue.createApp(app).mount('#container')
        }
    </script>
</head>
<body>
    <div id="container">
        <input type="button" value="动画切换" @click="toggle">
        <transition
        enter-active-class="animate_animated   animate_rotateIn"
        leave-active-class="animate_animated   animate_rotateOut">
            <div v-if="flag" class="box">结合Animate.css实现动画效果</div>
        </transition>
    </div>
</body>
</html>
```

点击"动画切换"按钮，div 旋转离开；再次点击"动画切换"按钮，div 旋转进入，其效果如图 5-12 所示。

图 5-12　旋转动画效果

5.3　群组对象添加过渡或动画

前面的过渡或者动画案例都是添加给单一对象的，如果要添加给群组对象，例如多个 \<li>\</li> 标签，则要使用 transition-group 作为父标签。transition-group 的子一级必须要有 :key 属性，否则页面就不能正常渲染。v-for 中使用 :key 属性的作用是来标识每一个循环出来的对象，便于在操作的时候能够找到该对象。

5.3.1 给工作计划表添加动画

给工作计划表添加动画

下面给工作计划表（参照【4.1 工作计划表】）添加过渡效果。当添加数据时，新数据从右侧进入；当删除数据时，对应的 li 向右侧移动删除。下面为案例代码。

CSS 中新增过渡效果的代码如下：

```css
*{ margin:0; padding: 0 }
body{background: #CCC;font-family: "微软雅黑"}
#plan{ width: 600px; background: #FFF; height: 700px; margin:0 auto; padding: 15px; position: relative; overflow: hidden; }    /* 给 #plan 添加溢出隐藏 */
#plan h1{ text-align: center; color: #F00; font-size: 40px; line-height: 60px}
#plan .txt{ width: 500px; height: 30px; border:1px #000 solid;outline: none }
#plan .btn{ width: 60px; height: 32px; margin-left: 5px; }
#plan h3{ line-height: 50px; }
#plan li{ list-style: none; padding-left: 20px; line-height: 35px; border-bottom: 1px #CCC dashed }
#plan li.on{ border:2px #C00 dashed; }
#plan li span{float: right; margin-right: 10px; color: #999; cursor: pointer;}
#plan li span:hover{ color: #000 }
#plan .footer{ position: absolute; left:15px; bottom: 20px; width: 600px;
    border-top:1px #CCC solid; line-height: 50px; }
#plan .footer .left{float: left;}
#plan .footer .right{ float: right; cursor: pointer; }

/* 添加过渡对应的 CSS */
.v-enter-from,.v-leave-to{ transform: translateX(500px); opacity: 0;}
.v-enter-active,.v-leave-active{ transition: all 0.5s;}
```

HTML 和 JS 代码如下：

```html
<!DOCTYPE html>
<html lang="en">
<head>
    <meta charset="UTF-8">
    <meta http-equiv="X-UA-Compatible" content="IE=edge">
    <meta name="viewport" content="width=device-width, initial-scale=1.0">
    <title>工作计划表添加动画</title>
    <link rel="stylesheet" href="./css/style.css">
    <script src="./js/vue3.js"></script>
    <script>
```

```
            window.onload=function(){
                const app={
                    data(){
                        return {
                            planList:[],
                            newPlan:",
                        }
                    },
                    methods:{
                        add(){
                            if(this.newPlan!="){
                                this.planList.push(this.newPlan)
                                this.newPlan="
                            }
                        },
                        del(i){
                            this.planList.splice(i,1)
                        },
                        delAll(){
                            this.planList=[]
                        },
                    },
                }
                Vue.createApp(app).mount('#container')
            }
        </script>
    </head>
    <body>
        <div id="container">
            <div id="plan">
                <h1>工作计划表</h1>
                <input type="text" class="txt" v-model="newPlan" @keydown.enter="add">
                <input type="button" class="btn" value="添加" @click="add">
                <h3>您目前的计划有：</h3>
                <transition-group tag="ul">
                    <li v-for="(value,index) in planList" :key="index">
                        {{index}}:{{value}}<span @click="del(index)">删除</span>
                    </li>
                </transition-group>
```

> transition-group 默认被渲染为 span 标签，使用 tag 属性设置渲染为 ul 标签

> 循环对象必须要有 :key 属性

```
                <div class="footer" v-if="planList.length>0">
                    <p class="left">计划任务共<strong>{{planList.length}}</strong>项</p>
                    <p class="right" @click="delAll()">删除全部</p>
                </div>
            </div>
        </div>
    </body>
</html>
```

添加一条新数据以后,新数据从右侧飞入进来,因为 #plan 添加了 overflow: hidden,所以超出 #plan 的部分被隐藏了,其效果如图 5-13 所示。

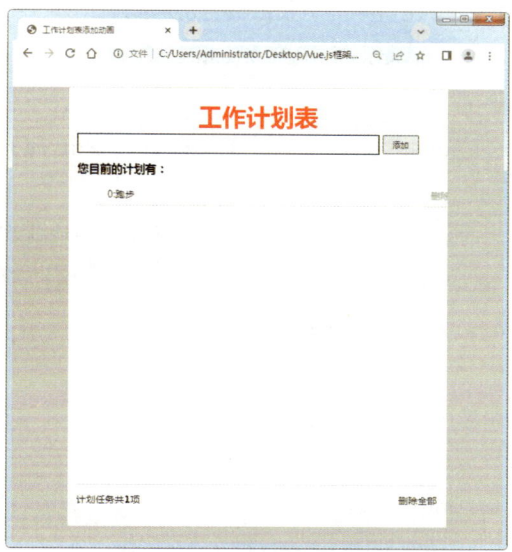

图 5-13　工作计划表添加动画

5.3.2　给焦点图添加过渡效果

下面案例给焦点图添加图片显示的过渡效果,让图片通过透明度的改变慢慢显示出来。焦点图原始代码请参看【4.3 焦点图效果】。焦点图添加过渡效果的代码如下:

```
<!DOCTYPE html>
<html lang="en">
<head>
    <meta charset="UTF-8">
    <meta http-equiv="X-UA-Compatible" content="IE=edge">
    <meta name="viewport" content="width=device-width, initial-scale=1.0">
    <title>焦点图效果</title>
    <style>
        *{ margin: 0; padding: 0;}
        .focus{ width: 590px; height: 320px; margin: 100px auto; border:2px red solid; position:relative;}
```

```css
.pic{ width: 590px; height: 320px; position:absolute}
.focus .pic img{ width: 590px; height: 320px;  }
.focus .pic h3{ position: absolute; left: 10px; bottom: 20px; color: #FFF; font-size: 18px;}
.focus .left,.focus .right{ width: 40px; height: 60px; background-color: rgba(0,0,0,0.5); color: #FFF; position: absolute; top:140px; font-size: 40px; line-height: 60px;  cursor: pointer;}
.focus .right{ right: 0;}
.focus .list{ width: 100px; height: 20px; position: absolute; right: 20px; bottom: 20px; }
.focus .list li{ width: 12px; height: 12px; background-color: rgb(255, 255, 255); border-radius: 50%; float: left; margin: 5px; list-style: none; cursor: pointer;}
.focus .list li.on{ background-color: rgb(43, 107, 226); transform: scale(1.3,1.3);}

.v-enter-from{ opacity: 0;}
.v-enter-active{ transition: all 1s;}
```

> 只添加进入的动画效果

```html
</style>
<script src="./js/vue3.js"></script>
<script>
    window.onload = function () {
      const app = {
        data() {
          return {
            focusPic:[
              {
                imgSrc:"./images/pic1.jpg",
                title:"中国的火星探测器"天问一号"成功实现火星探测"
              },
              {
                imgSrc:"./images/pic2.jpg",
                title:"中国的天文观测设备取得丰硕成果"
              },
              {
                imgSrc:"./images/pic3.jpg",
                title:"国产C919大型客机订单破千"
              },
            ],
```

```
                    :0
                }
            },
            methods:{
                prevPic(){
                    if(this.i==0){
                        this.i=this.focusPic.length-1
                    }
                    else{
                        this.i--
                    }
                },
                nextPic(){
                    if(this.i==this.focusPic.length-1){
                        this.i=0
                    }
                    else{
                        this.i++
                    }
                },
                startPic(){
                    this.timer = setInterval(()=>{
                        this.nextPic()
                    },2000)
                },
                stopPic(){
                    clearInterval(this.timer)
                }
            },
            created(){
                this.startPic()
            }
        }
        Vue.createApp(app).mount('.cont')
    }
    </script>
</head>
<body>
    <div class="cont">
```

```html
<div class="focus" @mouseover="stopPic" @mouseout="startPic">
    <div class="pic" v-for="(value,index) in focusPic" :key="index">
        <transition tag="div">
            <div class="item" v-if="i==index">
                <img :src="value.imgSrc">
                <h3>{{value.title}}</h3>
            </div>
        </transition>
    </div>
    <div class="left" @click="prevPic()">&lt;</div>
    <div class="right" @click="nextPic()">&gt;</div>
    <div class="list">
        <ul>
            <li v-for="(value,index) in focusPic" :class="{on:index==i}"
                @click="i=index">
            </li>
        </ul>
    </div>
</div>
</div>
</body>
</html>
```

> 因为 v-if 添加在 .item 对象上，所有在它的外边包含 transition，而不是 transition-group

案例执行的效果是图片可以使用透明度进行切换的一系列效果，如图 5-14~图 5-16 所示。

图 5-14　使用透明度过渡 1

图 5-15　使用透明度过渡 2

图 5-16　使用透明度过渡 3

课后练习题

1. Vue 过渡需要将对象包含在一对_____标签中，这实际上是 Vue 提供的封装组件。

2. Vue 过渡中的 Enter 阶段对应的三个类是_____、_____、_____，Leave 阶段对应的三个类是_____、_____、_____。

3. Vue 3.0 版本使用的是 .v-enter-from 和 .v-leave-from，而在 Vue 2.0 版本对应的类分别是_____

和_____。

4.如果页面中有多个对象需要添加过渡效果，但是每一个对象的过渡效果不同，则需要给 transition 添加一个_____属性。

5.如果要添加给群组对象，例如多个 <li></li> 标签，则要使用_____作为父标签。该标签必须要有_____属性，否则页面就不能正常渲染。

第 6 章

Axios 的使用

导言

"绿水青山就是金山银山",我国坚定不移地走生态优先、节约集约、绿色低碳的发展道路,着力推动经济社会发展全面绿色转型。近年来,我国在生态环境保护方面取得了显著的成就,森林覆盖率持续提升,湿地保护成效显著,大气质量持续得到改善。在本章中我们将学习使用 Axios 完成数据请求并且制作天气预报的综合案例。

学习内容

本章一共有 4 节。6.1 节主要讲解 Axios 的基本使用,其中涉及 Axios 的语法格式,有参数的情况下发起网络请求的格式。6.2 节讲解了"通过经纬获取地理信息"案例的制作方法。6.3 节讲解了"获取随机背景图"案例的制作方法。6.4 节通过一个综合案例"获取城市天气"的制作来提升发起网络请求的综合应用能力。

学习目标

1. 掌握 Axios 的引入方法。
2. 掌握 Axios 发起 GET 请求的基本语法格式。
3. 掌握 Axios 发起请求时带有参数的使用方法。
4. 能够根据需要正确选取所需要的数据。
5. 掌握对返回数据进行进一步处理的各种方法。

学习重点

1. 掌握 Axios 的引入和发起网络请求的基本格式。
2. 掌握 Axios 中带有参数请求的基本格式。
3. 能够分析项目需求,灵活选择和处理数据。

6.1 使用 Axios 发起网络请求

6.1.1 Axios 的基本用法

前面五个章节中，data 的数据都是由我们自己在编译器中输入的，那如何获取网络上的数据呢？这就要使用到 Axios。Axios 是一个基于 promise 的网络请求库。网络上的数据是通过 API 数据接口的形式获取的，可以使用 Axios 来发起请求获取 API 中的接口数据。

特别说明：本章节中的 API 数据接口均为从网络上收集的免费接口，仅为教学演示使用。

Axios 的网址为 https://www.axios-http.cn，其网站界面如图 6-1 所示，点击"起步"按钮可以查看 Axios 的使用文档，如图 6-2 所示。Axios 功能丰富，在其使用文档中对 Axios 的用例、POST 请求、请求配置、响应结构、默认配置、拦截器、错误处理、取消请求等多方面的使用进行了说明。本章节仅使用了 Axios 的基本用法。

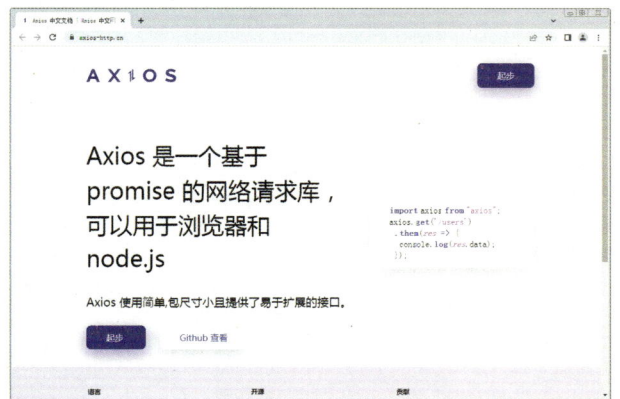

图 6-1　Axios 官网　　　　　　　　　图 6-2　Axios 的使用文档

点击 Axios 网站左侧的"用例"可以看到发起 GET 请求的格式，如图 6-3 所示。

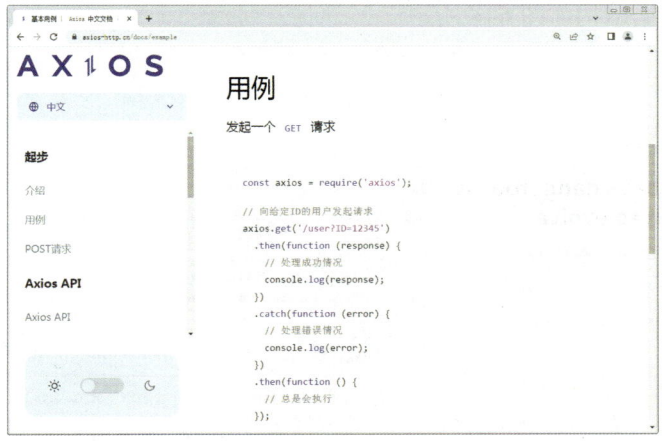

图 6-3　Axios 发起 GET 请求

GET 请求的基本格式为：

get（请求地址）.then（请求成功的回调函数）.catch（请求失败的回调函数）

也可以写成：

get（请求地址）.then（请求成功的回调函数，请求失败的回调函数）

例如，发起请求的代码为：

```
axios.get('https://autumnfish.cn/api/joke').then((response)=>{
    // 处理请求成功的情况
    console.log(response)
}).catch( (error)=> {
    // 处理请求失败的情况
    console.log(error);
})
```

请求的 API 接口地址

也可以写成：

```
axios.get('https://autumnfish.cn/api/joke').then((response)=>{
    // 处理请求成功的情况
    console.log(response)
},(error)=> {
    // 处理请求失败的情况
    console.log(error);
})
```

6.1.2　获取一句英文

下面案例是从 API 接口中获取一句英文。具体的接口配置如下：

请求地址：`https://api.oioweb.cn/api/common/OneDayEnglish`

请求方法：`get`

请求参数：无

请求成功的状态码：200

发送请求后，接口返回的数据是一个 JSON 对象，其中 data 是关键的数据，需要使用变量将所需的数据保存下来。返回的数据结构如图 6-4 所示。

获取一句英文

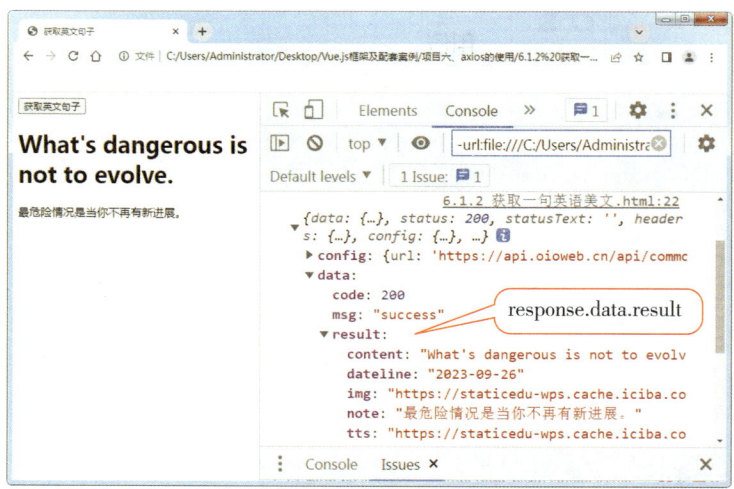

图 6-4　返回的数据

response.data.result 中有所需的英文以及翻译。将英文赋值给变量 Eng，将翻译赋值给变量 note，点

击"获取英文句子"按钮时将这两个变量渲染到页面,其代码如下:

```html
<!DOCTYPE html>
<html lang="en">
<head>
    <meta charset="UTF-8">
    <meta http-equiv="X-UA-Compatible" content="IE=edge">
    <meta name="viewport" content="width=device-width, initial-scale=1.0">
    <title>获取英文句子</title>
    <script src="./js/vue3.js"></script>
    <script src="./js/axios.min.js"></script>
    <script>
        window.onload=function(){
            const app={
                data(){
                    return{
                        Eng:"",
                        note:""
                    }
                },
                methods:{
                    getEng(){
axios.get('https://api.oioweb.cn/api/common/OneDayEnglish').then((response)=>{
                            console.log(response)
                            this.Eng=response.data.result.content
                            this.note=response.data.result.note
                        }).catch( (error)=> {
                            // 处理错误情况
                            console.log(error);
                        })
                    }
                }
                Vue.createApp(app).mount(".cont")
            }
    </script>
</head>
<body>
    <div class="cont">
```

先链接 vue.js,再链接 axios.min.js

将返回的数据分别赋值给变量 Eng 和变量 note

```
            <input type="button" value="获取英文句子" @click="getEng()">
            <h1>{{Eng}}</h1>
            <p>{{note}}</p>
        </div>
    </body>
</html>
```
将两个变量渲染到页面

点击"获取英文句子"按钮,页面中出现一句英文及其中文翻译,如图 6-5 所示。

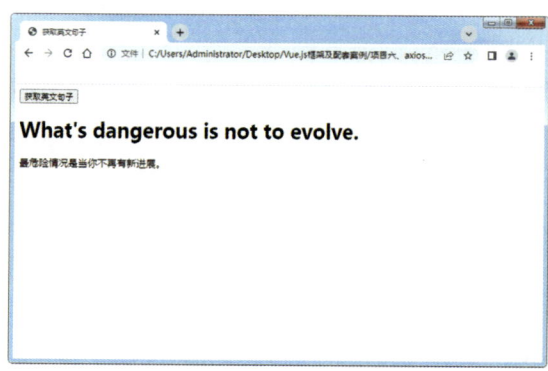

图 6-5　获取的英文句子及其中文翻译

6.1.3　获取垃圾分类

获取垃圾分类的接口配置如下:

请求地址:https://api.oioweb.cn/api/common/rubbish
请求方法:get
请求参数:name
参数说明:name 是垃圾的名称

有参数的请求格式为 get(请求地址?参数=参数值),当返回的数据是数组时,在 HTML 中需要使用 v-for 指令来对数组进行循环。案例代码如下:

```
<!DOCTYPE html>
<html lang="en">
<head>
    <meta charset="UTF-8">
    <meta http-equiv="X-UA-Compatible" content="IE=edge">
    <script src="./js/axios.min.js"></script>
    <script>
        window.onload=function(){
            const app={
                data(){
                    return {
                        fenlei:""
                    }
```

```
                },
            methods:{
                getFenlei(){
                    axios.get('https://api.oioweb.cn/api/common/rubbish?name=骨头').then(
                        (response)=>{
                            console.log(response)
                            this.fenlei=response.data.result
                        }).catch((err)=>{
                            console.log(err)
                        })
                    }
                }
            Vue.createApp(app).mount(".cont")
        }
    </script>
</head>
<body>
    <div class="cont">
        <input type="button" value="获取" @click="getFenlei()">

        <ul>
            <li v-for="item in fenlei">{{item.name}}——{{item.explain}}</li>
        </ul>
    </div>
</body>
</html>
```

> 使用 "name=骨头" 来传递参数

> response.data.result 获取多条垃圾分类的结果

> 使用 v-for 对多条数据进行循环

请求的返回值以及页面效果如图 6-6 所示。

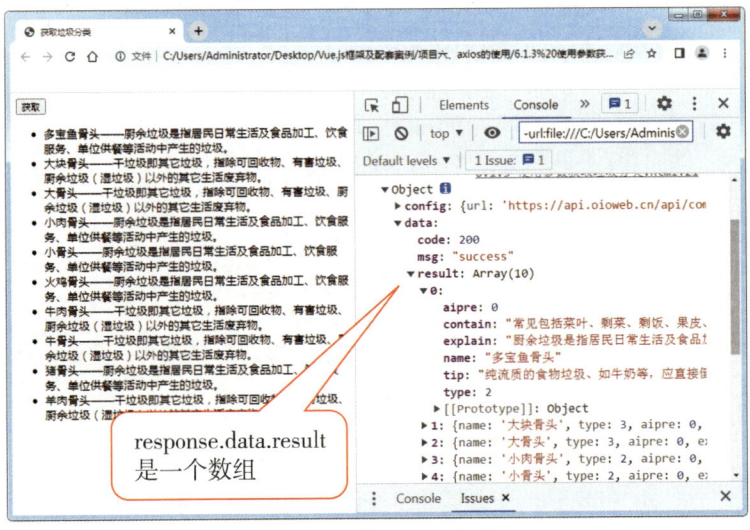

> response.data.result 是一个数组

图 6-6　获取垃圾分类

6.1.4　通过输入框传递参数

上面案例中的参数是直接使用的，下面将案例进行改进，通过输入框来传递参数。其设计思路是设置一个变量与文本框进行双向数据绑定，然后将该变量作为参数进行传递，具体代码如下：

```html
<!DOCTYPE html>
<html lang="en">
<head>
    <meta charset="UTF-8">
    <meta http-equiv="X-UA-Compatible" content="IE=edge">
    <meta name="viewport" content="width=device-width, initial-scale=1.0">
    <title>通过输入框传递参数</title>
    <script src="./js/vue3.js"></script>
    <script src="./js/axios.min.js"></script>
    <script>
        window.onload=function(){
            const app={
                data(){
                    return {
                        fenlei:"",
                        name:""
                    }
                },
                methods:{
                    getFenlei(){
                        axios.get('https://api.oioweb.cn/api/common/rubbish?name='+this.name).then(
                            (response)=>{
                                console.log(response.data)
                                this.fenlei=response.data.result
                            }).catch((error)=>{
                                console.log(error)
                            })
                    }
                }
            }
            Vue.createApp(app).mount(".cont")
        }
    </script>
</head>
<body>
    <div class="cont">
```

> this.name 是变量，不能在引号里面，要使用加号与前面的字符串连接起来

```
        输入名称:<input type="text" v-model="name">         与变量 name 进行双向数据绑定
         
        <input type="button" value="获取" @click="getFenlei()">
        <ul>
            <li v-for="item in fenlei">{{item.name}}——{{item.explain}}</li>
        </ul>
    </div>
</body>
</html>
```

案例效果如图6-7所示。在Axios官网上还提供了参数传递的其他方法，如图6-8所示。上述案例接口连接部分的代码可以改写为：

```
getFenlei(){
            axios.get('https://api.oioweb.cn/api/common/rubbish',{
                params:{
                    name:this.name                    将参数写在 params 中
                }
            }).then((response)=>{
                console.log(response.data)
                this.fenlei=response.data.result
            }).catch((error)=>{
                console.log(error)
                })
            }
```

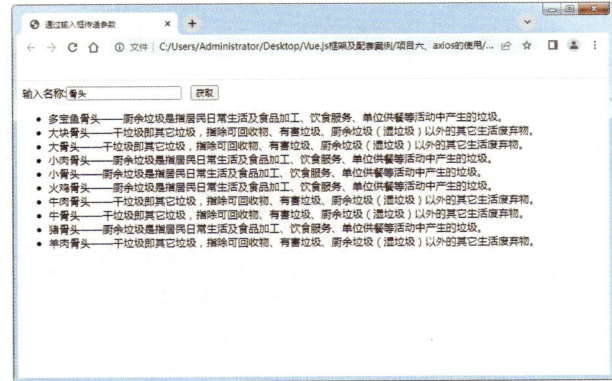

图6-7　使用文本框传递参数

图6-8　参数传递的第二种方法

6.2 通过经纬度数获取地理信息

6.2.1 项目描述

通过经纬度获取地理信息的接口配置如下：

请求地址：https://api.oioweb.cn/api/ip/geocoder
请求方法：get
请求参数：lng, lat
参数说明：lng 类型为数字，表示经度，正数表示东经，负数代表西经；
　　　　　lat 类型为数字，表示纬度，正数表示北纬，负数代表南纬；

要求在文本框中分别输入经度数值和纬度数值，点击"查看位置"按钮可以在下面显示查询出的详细地址信息，包括国家、省份、城市、地区、地址五个项目。例如，当经度输入115，纬度输入40时的效果如图6-9所示。

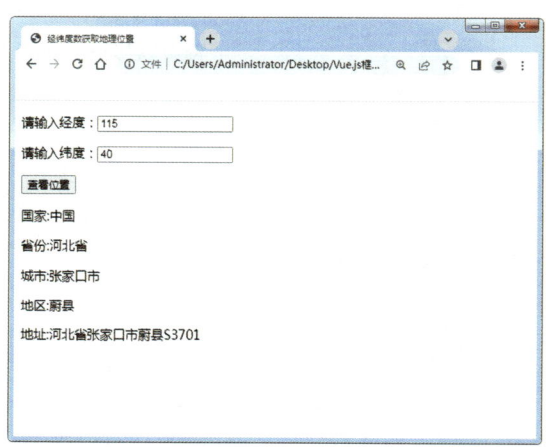

图6-9　项目效果

6.2.2 项目分析

该案例中发起请求时有两个参数，一个是经度，一个是纬度，可以在params中将两个参数传递过去。两个文本框要与经度和纬度的变量进行双向数据绑定。点击"查看位置"按钮时发起网络请求，对返回的数据进行处理，在页面中显示出对应的效果。

6.2.3 项目实施

项目实施步骤1：完成页面静态效果，发起网络请求

在data中设置两个变量：jing和wei，分别与两个文本框进行双向数据绑定。设置变量place用来保存返回数据中的地理信息，其代码如下：

```
<!DOCTYPE html>
<html lang="en">
```

```html
<head>
    <meta charset="UTF-8">
    <meta http-equiv="X-UA-Compatible" content="IE=edge">
    <meta name="viewport" content="width=device-width, initial-scale=1.0">
    <title>经纬度数获取地理位置</title>
    <script src="./js/vue3.js"></script>
    <script src="./js/axios.min.js"></script>
    <script>
        window.onload=function(){
            const app={
                data(){
                    return {
                        jing:"",
                        wei:"",
                        place:""
                    }
                },
                methods:{
                    getPlace(){
                        axios.get('https://api.oioweb.cn/api/ip/geocoder',{
                            params:{
                                lng:this.jing,
                                lat:this.wei,
                            }}).then((response)=>{
                                console.log(response)
                                this.place=response.data.result
                            },(error)=>{
                                console.log(error)
                            })
                    }
                }
            }
            Vue.createApp(app).mount(".cont")
        }
    </script>
</head>
<body>
    <div class="cont">
        <p>请输入经度：<input type="text" v-model="jing"></p>
```

> 传递两个参数

> 参照图 6-10，地理的详细信息通过 response.data.result 获取，获取后保存到变量 place 中

```
            <p>请输入纬度：<input type="text" v-model="wei"></p>
            <input type="button" value="查看位置" @click="getPlace()">
        </div>
    </body>
</html>
```

页面的静态界面如图 6-10 所示，在经度文本框中输入 115，在纬度文本框中输入 40，点击"查看位置"后，页面加载的数据如图 6-11 所示，详细的地理信息通过 response.data.result 获取。

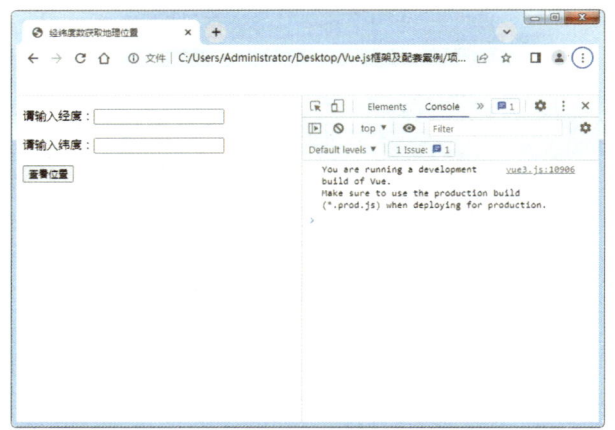

图 6-10 经纬度数静态界面

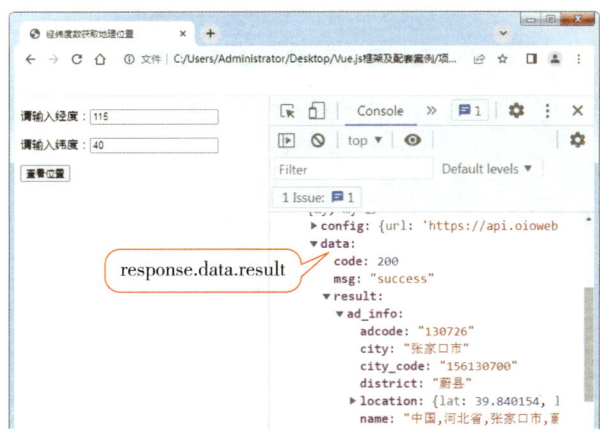

图 6-11 地理信息的数据结构

项目实施步骤 2：将数据渲染到页面中

在页面中需要呈现出来的内容有国家、省份、城市、地区和地址。观察返回数据的结构，如图 6-12 所示，可知国家对应的值为 response.data.result.ad_info.nation。

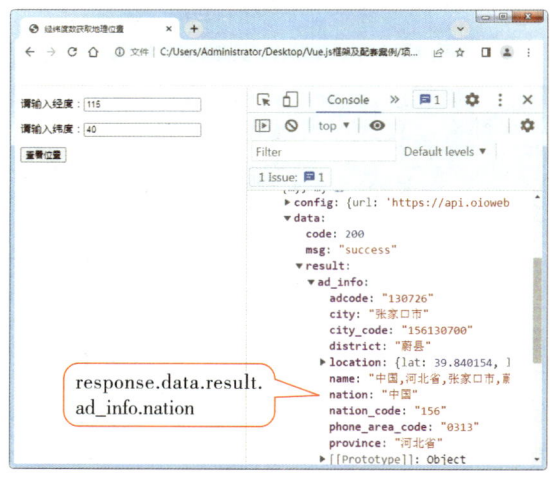

图 6-12 经纬获取地理位置发起请求

在 HTML 中对变量进行渲染，其代码如下：

```
<!DOCTYPE html>
<html lang="en">
<head>
    <meta charset="UTF-8">
    <meta http-equiv="X-UA-Compatible" content="IE=edge">
```

```html
<meta name="viewport" content="width=device-width, initial-scale=1.0">
<title>经纬度数获取地理位置</title>
<script src="./js/vue3.js"></script>
<script src="./js/axios.min.js"></script>
<script>
    window.onload=function(){
        const app={
            data(){
                return {
                    jing:"",
                    wei:"",
                    place:""
                }
            },
            methods:{
                getPlace(){
                    axios.get('https://api.oioweb.cn/api/ip/geocoder',{
                        params:{
                            lng:this.jing,
                            lat:this.wei,
                        }}).then((response)=>{
                            console.log(response)
                            this.place=response.data.result
                        },(error)=>{
                            console.log(error)
                        })
                }
            }
        }
        Vue.createApp(app).mount(".cont")
    }
</script>
</head>
<body>
    <div class="cont">
        <p>请输入经度：<input type="text" v-model="jing"></p>
        <p>请输入纬度：<input type="text" v-model="wei"></p>
        <input type="button" value="查看位置" @click="getPlace()">
        <div class="address">
```

```
            <p>国家:{{place.ad_info.nation}}</p>
        </div>
    </div>
</body>
</html>
```

> 在页面中渲染出国家所对应的值

页面加载以后报错，效果如图 6-13 所示。如果将代码修改成 "<p> 国家 :{{place.ad_info}}</p>"，再加载页面则没有报错，在文本框中分别输入经纬度数，页面效果如图 6-14 所示。这其中的原因是刚加载页面的时候，place 的初始值为一个空对象，使用 {{place.ad_info}} 对应的值是 undefined，这时是不会报错的，但是 {{place.ad_info.nation}} 相当于是 undefined.nation 就会报错。

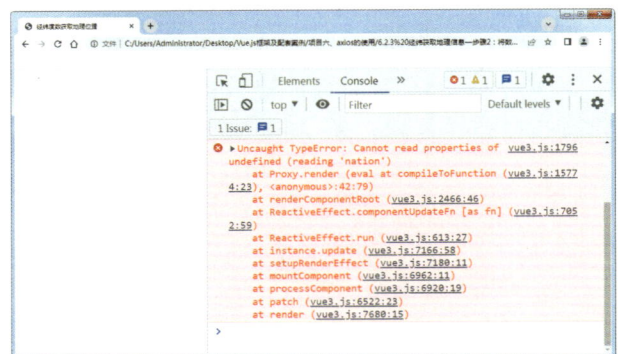
图 6-13 页面报错

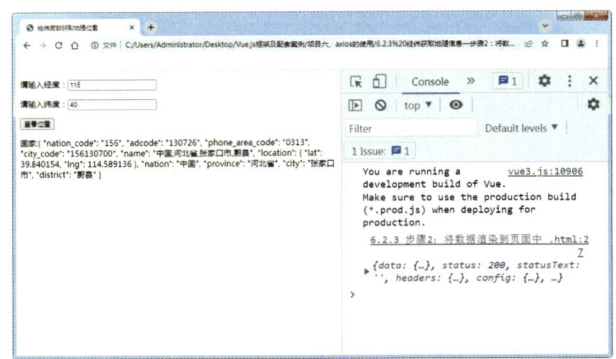
图 6-14 place.ad_info 加载成功

项目实施步骤 3：完成地理信息获取

修改步骤 2 中的代码，在 data 中设置变量 nation、province、city、district、address，依次用于保存国家、省份、城市、区域和地址，在发起请求成功以后对这个变量进行赋值。其代码如下：

```
<!DOCTYPE html>
<html lang="en">
<head>
    <meta charset="UTF-8">
    <meta http-equiv="X-UA-Compatible" content="IE=edge">
    <meta name="viewport" content="width=device-width, initial-scale=1.0">
    <title>经纬度数获取地理位置</title>
    <script src="./js/vue3.js"></script>
    <script src="./js/axios.min.js"></script>
    <script>
        window.onload=function(){
            const app={
                data(){
                    return {
                        jing:"",
                        wei:"",
                        nation:"",// 国家
```

```
                    province:"", //省份
                    city:"", //城市
                    district:"", //地区
                    address:"", //地址
                
                }
            },
            methods:{
                getPlace(){
                    axios.get(´https://api.oioweb.cn/api/ip/geocoder´,{
                        params:{
                            lng:this.jing,
                            lat:this.wei,
                        }}).then((response)=>{
                        console.log(response)
                        this.nation=response.data.result.ad_info.nation
                        this.province=response.data.result.ad_info.province
                        this.city=response.data.result.ad_info.city
                        this.district=response.data.result.ad_info.district
                        this.address=response.data.result.address
                    
                    },(error)=>{
                        console.log(error)
                    })
                }
            }
            Vue.createApp(app).mount(".cont")
        }
    </script>
</head>
<body>
    <div class="cont">
        <p>请输入经度：<input type="text" v-model="jing"></p>
        <p>请输入纬度：<input type="text" v-model="wei"></p>
        <input type="button" value="查看位置" @click="getPlace()">
        <div class="address">
            <p v-if="nation">国家:{{nation}}</p>
            <p v-if="province">省份:{{province}}</p>
```

> 发起请求成功以后对各个变量进行赋值

> 使用v-if优化页面，当变量的值为空值时对应的p标签不显示

```
                <p v-if="city">城市:{{city}}</p>
                <p v-if="district">地区:{{district}}</p>
                <p v-if="address">地址:{{address}}</p>
            </div>
        </div>
    </body>
</html>
```

当在页面输入经度115,纬度40时,页面的效果如图6-15所示;当输入经度-115,纬度40时,页面的效果如图6-16所示。

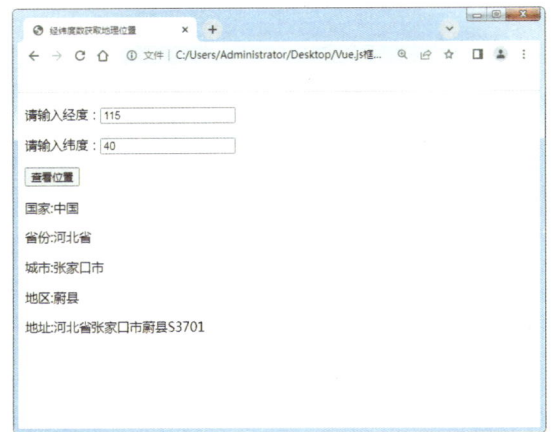

图6-15 查询(115,40)的效果

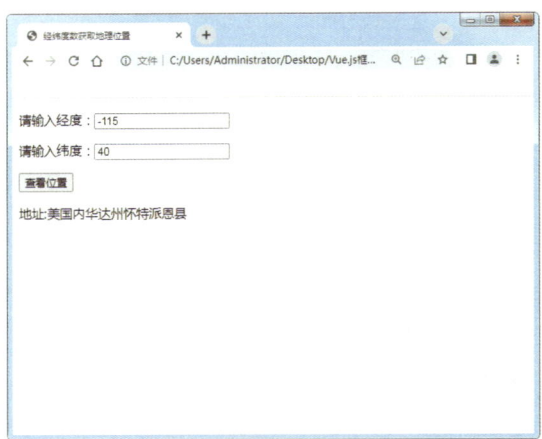

图6-16 查询(-115,40)的效果

6.3 获取随机背景图

6.3.1 项目描述

通过API数据接口发起请求,从网络获取图片作为背景图片,接口提供的图片一共有8张,将这8张图片做成小缩略图,点击可以切换背景图片。具体效果如图6-17所示。

图6-17 切换背景图片

获取随机背景图的接口配置如下：

请求地址：https://api.oioweb.cn/api/bing

请求方法：get

请求参数：无

6.3.2 项目分析

该项目中将 8 张图片渲染到页面中要使用 v-for 指令。设置一个变量来记录目前选中图片的索引值，当选中图片时，能够出现白色的边框线，使用的是类的绑定。

6.3.3 项目实施

项目实施步骤 1: 完成静态界面

```html
<!DOCTYPE html>
<html lang="en">
<head>
    <meta charset="UTF-8">
    <meta http-equiv="X-UA-Compatible" content="IE=edge">
    <meta name="viewport" content="width=device-width, initial-scale=1.0">
    <title>切换背景图</title>
    <style>
        *{ margin: 0; padding: 0;}
        .lst{ width: 500px; height: 80px; position: absolute; right: 0; top: 0;}
        .lst li{ list-style: none; float: left; width: 50px; margin: 3px; height: 30px; cursor: pointer;}
        .lst li.on{border: 1px #999999 solid;}
        .lst li img{ width: 50px; height: 30px;}
        .bg{ height: 100vh; background-size: cover;}
    </style>
</head>
<body>
    <div class="cont" >
        <div class="lst">
            <ul>
                <li class="on"><img src="./bg.jpg"></li>   ← 第一个 li 添加了 on 类
                <li><img src="./bg.jpg"></li>
                <li><img src="./bg.jpg"></li>
                <li><img src="./bg.jpg"></li>
                <li><img src="./bg.jpg"></li>
                <li><img src="./bg.jpg"></li>
```

```
                <li><img src="./bg.jpg"></li>
                <li><img src="./bg.jpg"></li>
            </ul>
        </div>
        <div class="bg" style="background-image:url(./bg.jpg);"></div>
    </div>
</body>
</html>
```

背景切换的静态界面效果如图 6-18 所示。

项目实施步骤 2：
发起请求，完成背景切换

图 6-18　背景切换静态界面

项目实施步骤 2：发起请求，完成背景切换

设置还是 getPic( ) 用于发起网络请求，因为页面一加载的时候背景图片就要出现，所以在生命周期函数 created 中调用 getPic( ) 函数。设置变量 i 用于记录正在显示图片的索引值，点击小缩略图，可以改变 i 的值，具体代码如下：

```
<!DOCTYPE html>
<html lang="en">
<head>
    <meta charset="UTF-8">
    <meta http-equiv="X-UA-Compatible" content="IE=edge">
    <meta name="viewport" content="width=device-width, initial-scale=1.0">
    <title>切换背景图</title>
    <style>
        *{ margin: 0; padding: 0;}
        .lst{ width: 500px; height: 80px; position: absolute; right: 0; top: 0;}
        .lst li{ list-style: none; float: left; width: 50px; margin: 3px; height: 30px; cursor: pointer;}
        .lst li.on{border: 1px white solid;}
        .lst li img{ width: 50px; height: 30px;}
        .bg{ height: 100vh; background-size: cover;}
    </style>
```

```
<script src="./js/vue3.js"></script>
<script src="./js/axios.min.js"></script>
<script>
    window.onload=function(){
        const app={
            data(){
                return {
                    pic:"",
                    i:0
                }
            },
            methods:{
                getPic(){
                    axios.get('https://api.oioweb.cn/api/bing').then((response)=>{
                        console.log(response)
                        this.pic=response.data.result
                    },(err)=>{
                        console.log(err)
                    })
                }
            },
            created(){
                this.getPic()
            }
        }
        Vue.createApp(app).mount(".cont")
    }
</script>
</head>
<body>
    <div class="cont" >
        <div class="lst">
            <ul>
                <li v-for="(value,index) in pic":class="{on:i==index}" @click="i=index">
                    <img :src="value.url" alt="">
                </li>
            </ul>
```

> 动态绑定 on 类

```
            </div>
            <div class="bg" :style="{'background-image':'url('+pic[i].url+')'}"></div>
        </div>
    </body>
</html>
```

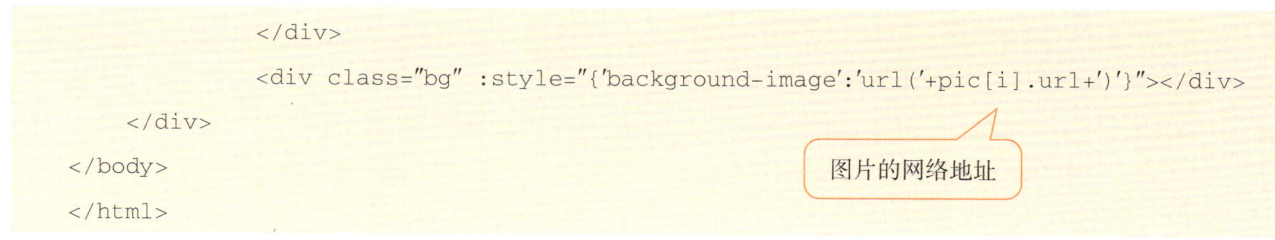

图片的网络地址

发起请求后返回的数据如图 6-19 所示。

图 6-19　发起请求获取数据

点击第 2 张图片的效果如图 6-20 所示，点击第 4 张图片的效果如图 6-21 所示。

图 6-20　点击第 2 张图的效果

图 6-21　点击第 4 张图的效果

6.4　获取城市天气

6.4.1　项目描述

本项目需要发起网络请求，获取当前城市的天气情况，其效果如图 6-22 所示。在页面中展现的数据包括当前的气温、天气提示、风向、风力、温度范围、空气湿度以及后面五天的天气预报等。页面显示的数据与请求到的数据并不完全一致，需要进行数据的处理，具体需要处理的数据以图 6-22 为例有：

（1）城市获取的数据是"济南市"，要处理成"济南"；
（2）天气提示的文字可能比较多，要截取前 10 个字符，后面连接上"……"；
（3）天气预报数组中一共有 7 条数据，只截取后五条数据进行循环；

（4）获取的时间格式为"2023-09-27"，要处理成"09/27"；

（5）请求到的数据中只有天气情况，要使用对应的天气图标来显示；

（6）头部的背景图能够根据天气情况进行切换，背景类型有晴天背景、多云背景、阴天背景、雨天背景和默认背景；

（7）湿度部分的背景颜色能够根据湿度进行改变，湿度低于50时的背景色为绿色，湿度为50~80时的背景色为橙色，湿度高于80时的背景色为红色，例如图6-22中的湿度为96，所以背景色为红色。

图6-22　获取天气情况

6.4.2　前置知识——字符串的相关处理方法

字符串的分割使用split方法。split方法将某个字符作为分割符，将一个字符串分割为多个部分存放在数组中。例如，下面代码：

```
<script>
    var s1="2023-09-25"
    var newArr=s1.split("-")
    console.log(newArr)//[2023,09,25]
    console.log(newArr[1]+"/"+newArr[2])// 结果为：09/25
</script>
```

取数组的第1位和第2位进行连接

对s1字符串使用"-"进行分割，分割成"2023"、"09"和"25"三部分。分割以后的各个部分存放在数组中。

字符串的替换使用replace方法。rplace方法可以将一个字符替换成另外一个字符，例如下面的代码将"市"替换为无。

```
<script>
    var s1="济南市"
    var s2=s1.replace('市','')
    console.log(s2)// 济南
</script>
```

将"市"替换为无，替换以后s2为"济南"

字符串的截取使用slice方法，其语法格式为slice（开始位置的索引值，结束位置的索引值）。例如：

```
<script>
    var s1="Vue.js"
    var s2=s1.slice(0,3)// 截取字符串中的第0,1,2位字符
    console.log(s2)// 新字符串s2的值为"Vue"
</script>
```

上面案例中使用slice方法截取s1字符串中索引值0~3之间的子串，这里索引值为3的那个字符是取不到的。所以会截取字符串中的第0，1，2位字符。slice除了可以截取字符串也可以来截取数组。例如：

```
<script>
    var arr1=[0,1,2,3]
    var arr2=arr1.slice(1,4)// 新数组arr2的值为[1,2,3]
    var arr3=arr1.slice(1,)// 新数组arr3的值为[1,2,3]
    console.log(arr2)
    console.log(arr3)
</script>
```

slice(1,)的作用是从索引值为1的位置一直截取到最后。

6.4.3　项目分析

获取接口数据以后要使用split、slice、replace等方法对数据进行处理。根据天气情况的id值构建图片路径的数组，让图片的索引值与id值相对应。背景图片和背景颜色的动态切换使用类的绑定来进行实现。

6.4.4　项目实施

项目实施步骤1：完成静态界面效果

在样式中设定多个类来设定不同的背景图片和背景颜色，头部的背景图使用默认情况下的背景图片，其代码如下：

```
<!DOCTYPE html>
<html lang="en">
<head>
    <meta charset="UTF-8">
    <meta http-equiv="X-UA-Compatible" content="IE=edge">
    <meta name="viewport" content="width=device-width, initial-scale=1.0">
    <title>城市天气</title>
    <link rel="stylesheet" href="./css/style.css">
</head>
<body>
    <div class="cont">
        <div class="top">
```

```html
<div class="head">
    <h3>北京天气 <span>2023-9-20</span></h3>
    <div class="tip">
        <i></i>
        <p>天气提示 </p>
    </div>
</div>

<div class="details">
    <h1>25℃ </h1>
    <div class="more">
        <h3>晴 东风 5 级 </h3>
        <h3>18℃ ~26℃ <span class="shidu green">50</span></h3>
    </div>
</div>
</div>
<div class="yubao">
    <h2>五天天气预报 </h2>
    <ul>
        <li>
            <p>9/20</p>
            <div class="pic">
                <img src="./images/0.png">
                <p>晴 </p>
            </div>
            <p>18℃ ~26℃ </p>
        </li>
        <li>
            <p>9/20</p>
            <div class="pic">
                <img src="./images/0.png">
                <p>晴 </p>
            </div>
            <p>18℃ ~26℃ </p>
        </li>
        <li>
            <p>9/20</p>
            <div class="pic">
                <img src="./images/0.png">
```

```html
                    <p>晴</p>
                </div>
                <p>18℃~26℃</p>
            </li>
            <li>
                <p>9/20</p>
                <div class="pic">
                    <img src="./images/0.png">
                    <p>晴</p>
                </div>
                <p>18℃~26℃</p>
            </li>
            <li>
                <p>9/20</p>
                <div class="pic">
                    <img src="./images/0.png">
                    <p>晴</p>
                </div>
                <p>18℃~26℃</p>
            </li>
        </ul>
    </div>
  </div>
</body>
</html>
```

CSS 代码如下：

```css
*{ margin: 0; padding: 0;}
body{background-color: lightblue;}
.cont{ width: 600px; height: 600px; margin: 0px auto; background-color:#FFFFFF; }
.top{ height: 160px; background-color: blueviolet;padding: 10px;background-image: url(../images/bg5.jpg);}
.sunny{ background-image: url(../images/bg1.jpg);}
.cloudy{background-image: url(../images/bg2.jpg);}
.overcast{background-image: url(../images/bg3.jpg);}
.rainy{background-image: url(../images/bg4.jpg);}
.cont .head{ height: 30px; padding: 15px 0 ;}
.cont .head h3{ float: left; font-size: 18px; color: #FFFFFF;}
.cont .head h3 span{ font-weight: normal; font-size: 12px; margin-left: 10px;}
```

> 多个背景图的类，分别对应晴天、多云、阴天和雨天

```css
.cont .head .tip{ float: right; height: 30px; width: 300px; margin-right: 30px; border-radius: 15px; background-color: rgba(0,0,0,0.5); color: #FFFFFF; line-height: 30px;}
.cont .head .tip i{ width: 10px; height: 10px; background-color: rgb(125, 238, 150); border-radius: 50%; float: left; margin-top: 10px; margin-left: 15px; margin-right: 15px;}
.cont .details{ height: 100px;}
.cont .details h1{ float: left; font-size: 60px; color: #FFFFFF;}
.cont .details .more{float: left; margin-left: 20px; line-height: 30px; margin-top: 10px; color: #333333;}
.cont .details .more .shidu{ display: inline-block; padding:0 5px; border-radius: 3px; color: #FFFFFF;}
.green{ background-color: limegreen;}
.orange{ background-color: orange;}
.red{ background-color: red;}
.yubao{padding-left: 20px;}
.yubao h2{ margin-bottom: 20px; margin-top: 25px;}
.yubao li{ list-style: none; float: left; width: 100px; text-align: center; margin: 5px; height: 150px; background-color: rgb(240, 240, 240); padding: 15px 0; border-radius: 10px;}
.yubao li .pic img{ width: 70px; height: 70px;}
.yubao li .pic p{ margin-top: -10px;}
```

> 多个背景颜色的类，分别对应绿色、橙色和红色

获取城市天气的静态界面效果如图 6-23 所示。

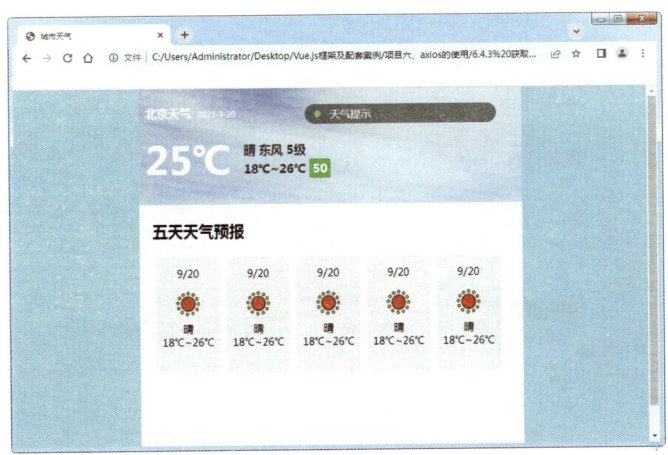

图 6-23 获取城市天气静态界面效果

项目实施步骤 2：发起网络请求获取数据

获取城市天气的接口配置如下：

请求地址：https://api.oioweb.cn/api/weather/GetWeather

请求方法：get

请求参数：无

script 代码为：

```
<script>
    window.onload=function(){
        const app={
            data(){
                return {
                }
            },
            methods:{
                getWeather(){
                    axios.get('https://api.oioweb.cn/api/weather/GetWeather').then(
                        (response)=>{
                            console.log(response)
                        },(err)=>{
                            console.log(err)
                        })
                }
            },
            created(){
                this.getWeather()
            }
        }
        Vue.createApp(app).mount('.cont')
    }
</script>
```

返回数据如图 6-24~ 图 6-26 所示。

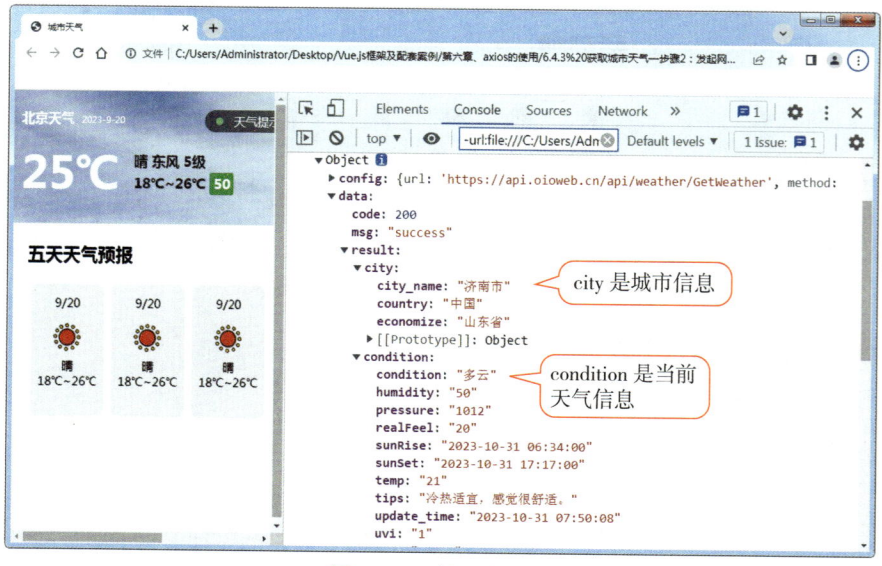

图 6-24　接口数据 1

第 6 章　Axios 的使用

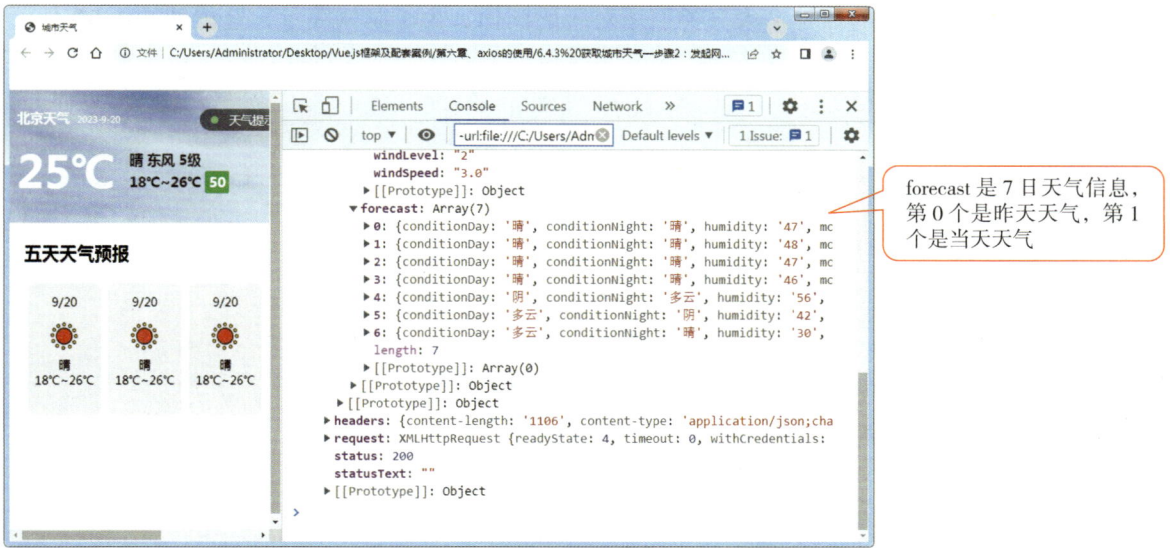

图 6-25　接口数据 2

forecast 是 7 日天气信息，第 0 个是昨天天气，第 1 个是当天天气

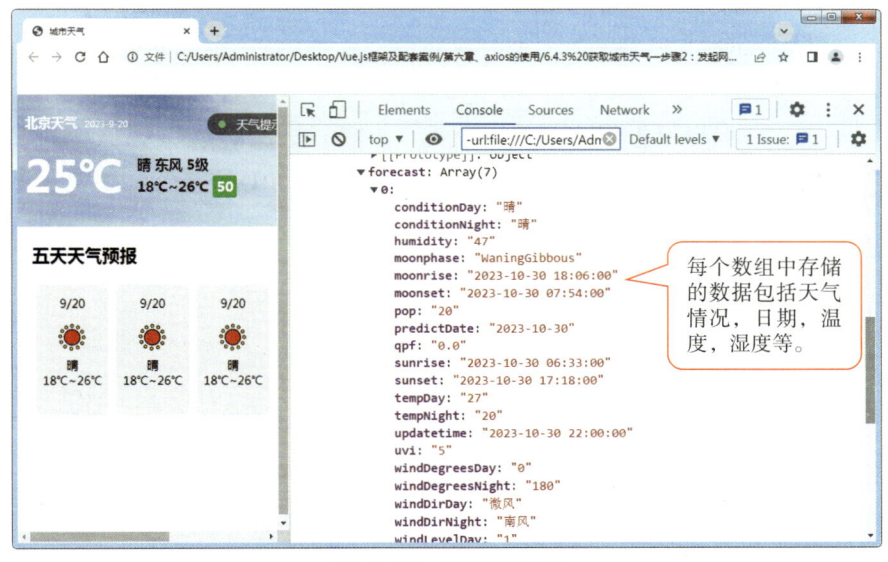

图 6-26　接口数据 3

每个数组中存储的数据包括天气情况，日期，温度，湿度等。

项目实施步骤 3：完成获取城市天气

city.city_name 的值是"济南市"，使用 replace 方法变成"济南"。condition.tips 是天气提示，使用 slice 方法截取 0~10 个字符。forecast 数组的长度是 7，使用 slice 方法截取后 5 天的数据。当天日期的格式是"2023-09-27"，使用 split 方法处理为"09/27"。图标图片存储在数组 icon 中，一共使用了 10 张图片，如图 6-27 所示，依次对应晴天、多云、阴天、阵雨等天气。背景图一共有 5 张，分别是晴天、多云、阴天、雨天和默认背景图片。icon 的获取是构造一个天气情况与图标的对应数据，天气的图标如图 6-27 所示。通过 findIicon 函数将天

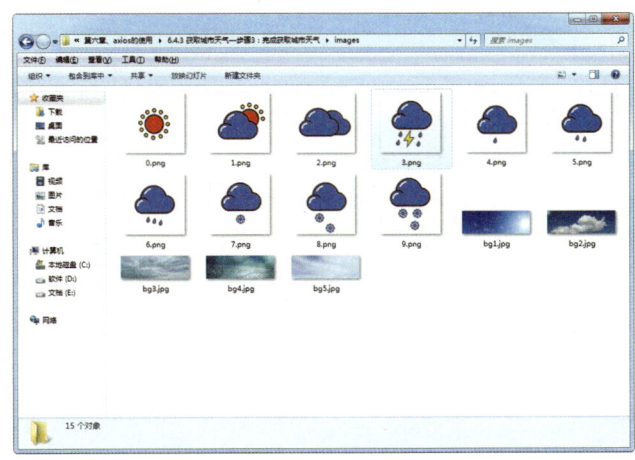

图 6-27　天气图标展示

135

气情况作为参数输入，返回的值是该天气类型所对应的图标，为了案例演示，这里列举了10种天气所对应的图标。具体代码如下：

```html
<!DOCTYPE html>
<html lang="en">
<head>
    <meta charset="UTF-8">
    <meta http-equiv="X-UA-Compatible" content="IE=edge">
    <meta name="viewport" content="width=device-width, initial-scale=1.0">
    <title>城市天气</title>
    <link rel="stylesheet" href="./css/style.css">
    <script src="./js/vue3.js"></script>
    <script src="./js/axios.min.js"></script>
    <script>
        window.onload=function(){
            const app={
                data(){
                    return {
                        city:"",
                        temp:"",
                        condition:"",
                        windDir:"",
                        windLevel:"",
                        forecast:"",
                        humidity:"",
                        tips:"",
                        tempNight:"",
                        tempDay:"",
                        now:"" ,
                        weatherIcon:[
                            {con:"晴",icon:'./images/0.png'},
                            {con:"多云",icon:'./images/1.png'},
                            {con:"阴",icon:'./images/2.png'},
                            {con:"雷阵雨",icon:'./images/3.png'},
                            {con:"小雨",icon:'./images/4.png'},
                            {con:"中雨",icon:'./images/5.png'},
                            {con:"大雨",icon:'./images/6.png'},
                            {con:"小雪",icon:'./images/7.png'},
                            {con:"中雪",icon:'./images/8.png'},
```

> 构造天气情况与图标的对应关系

```
                {con:"大雪",icon:'./images/9.png'},
            ],
        }
    },
    methods:{
        getWeather(){
            axios.get('https://api.oioweb.cn/api/weather/GetWeather').
             then((response)=>{
                console.log(response)
                weather=response.data.result
                this.city=weather.city.city_name.replace('市','')//城市
                this.temp=weather.condition.temp//温度
                this.condition=weather.condition.condition//天气情况
                this.windDir=weather.condition.windDir//风向
                this.windLevel=weather.condition.windLevel//风力
                this.forecast=weather.forecast//7天天气预报的整个数组
                this.humidity=weather.condition.humidity//湿度
                this.tips=weather.condition.tips  //天气提示
                this.tempNight=weather.forecast[1].tempNight//低温
                this.tempDay=weather.forecast[1].tempDay//高温
                this.now=weather.forecast[1].predictDate//当前日期
            },(err)=>{
                console.log(err)
            })
        },

        findIcon(con){
          var i=this.weatherIcon.findIndex(item=>{
                return item.con==con        ← 查找函数参数的索引值
            })
            return this.weatherIcon[i].icon  ← 返回天气所对应的图标
        }
    },
    created(){
        this.getWeather()
    }
}
Vue.createApp(app).mount('.cont')
```

```html
        }
    </script>
</head>
<body>
    <div class="cont">
        <div class="top"
        :class="{sunny:condition=='晴',cloudy:condition=='多云',
        overcast:condition=='阴',rainy:condition.indexOf('雨')!=-1}">
            <div class="head">
                <h3>{{city}}天气 <span>{{now}}</span></h3>
                <div class="tip">
                    <i></i>
                    <p>{{tips.length>=10?tips.slice(0,10)+'......':tips}}</p>
                </div>
            </div>


            <div class="details">
                <h1>{{temp}}℃ </h1>
                <div class="more">
                    <h3>{{condition}} {{windDir}} {{windLevel}}级 </h3>
                    <h3>{{tempNight}}℃ ~{{tempDay}}℃
                        <span class="shidu"
                        :class="{red:humidity>=80,orange:humidity>=50&&humidity<80,
                        :green:humidity<50}">{{humidity}}</span>
                    </h3>
                </div>
            </div>
        </div>
        <div class="yubao">
            <h2>五天天气预报</h2>
            <ul>

                <li v-for="item in forecast.slice(2,7)">
                    <p>{{item.predictDate.split('-')[1]}}/{{item.predictDate.split('-')[2]}}</p>
                    <div class="pic">
                        <img :src="findIcon(item.conditionDay)">
                        <p>{{item.conditionDay}}</p>
                    </div>
```

> 根据 condition 的值动态绑定背景图片所对应的类。如果 conditon 中包含"雨"字，使用 rainy 类

> 如果 tips 的长度超过 10，则截取前 10 个字符，否则不截取

> 湿度超过 80 时，使用红色背景色；湿度为 50~80 时，使用橙色背景色；湿度小于 50 时，使用绿色背景色

> 循环 forecast 的后 5 个数据

> 对时间进行截取

```
                    <p>{{item.tempNight}}℃~{{item.tempDay}}℃</p>
                </li>
            </ul>
        </div>
    </div>
</body>
</html>
```

雨天的页面效果如图 6-28 所示。阴天的页面效果如图 6-29 所示。雾天的页面效果如图 6-30 所示。

图 6-28　雨天页面效果

图 6-29　阴天页面效果

图 6-30　雾天页面效果

课后练习题

1. Axios 中使用 GET 请求发送网络请求获取 API 接口数据的基本格式为：get（_____）.then（_____）.catch（_____）

2. 有参数的请求格式为 get（_____），也可以将参数写在_____中。

3. split、slice、replace 中字符串的分割使用_____方法，字符串的替换使用_____方法，字符串的截取使用_____方法。

第 7 章

Vue.js 组件

导言

我国坚持新发展理念，经济踏浪前行，迈上高质量、高效、公平、可持续、安全的发展之路。经济结构进一步优化，高技术制造业、装备制造业和数字经济等产业不断壮大。人民生活水平持续提高，脱贫攻坚任务胜利完成，人民群众获得感显著增强。在本章我们将学习 Vue 组件的相关知识，并且完成制作购物车组件的综合案例。让我们学好专业知识，发挥自身特长为经济发展注入活力和动力。

学习内容

本章一共有 5 节。7.1 节主要讲解组件的基本应用，其中涉及的知识点包括组件的基本结构、Vue 2.0 组件的结构、组件的命名等。7.2 节主要讲解组件切换的方法。7.3 节主要讲解组件的通信，这是本章的重点，主要涉及父组件向子组件传递数据的方法和子组件向父组件传递数据的方法。7.4 节主要讲解插槽的多种类型及使用方法。7.5 节通过综合案例"购物车"组件来加深对组件的理解，提升组件的应用能力。

学习目标

1. 理解使用组件的意义。
2. 掌握组件注册以及组件应用的基本语法格式。
3. 掌握父子组件通信的实现方法。
4. 掌握插槽的意义以及使用方法。
5. 能够灵活应用组件来完成相应的功能。

学习重点

1. 掌握组件注册以及组件应用的基本语法格式。
2. 掌握父组件向子组件传递数据的方法。
3. 掌握子组件向父组件传递数据的方法。

7.1　组件的基本结构

创建 Vue 3.0 组件

7.1.1　创建 Vue 3.0 组件

组件（component）是 Vue.js 最强大的功能之一。组件是一个 HTML、CSS、JS 等的聚合体，这个聚合体是封装起来可重用的代码。组件化的设计理念可以将一个完整功能的项目分割成多个功能模块，有利于加快项目的进度和进行项目的复用。

Vue 中常用的组件有全局组件和局部组件，全局组件可以在任何组件模板中使用。局部组件在父组件中的"components"选项中进行注册。下面主要介绍全局组件的使用方法。需要注意的是 Vue 3.0 和 Vue 2.0 中组件的使用方法有所区别。

组件的本质是一个对象，这个对象中有 template（组件的模板结构）、data、computed 和 methods 等多个部分。组件要先注册然后使用。下面案例定义了一个组件，其代码如下：

```
<!DOCTYPE html>
<html lang="en">
<head>
    <meta charset="UTF-8">
    <meta http-equiv="X-UA-Compatible" content="IE=edge">
    <meta name="viewport" content="width=device-width, initial-scale=1.0">
    <title>Document</title>
    <script src="./js/vue3.js"></script>
    <script>
        window.onload = function () {
            const app = Vue.createApp({})// 创建一个Vue应用
            app.component('hello', {        // 全局注册一个名称为 hello 的组件
                template:
                    "<h1>组件</h1><h2 @click='sayHello()'>点击出现弹窗</h2>",
                data() {
                    return {
                        msg: "hello,vue!"
                    }
                },
                methods: {
                    sayHello() {
                        alert(this.msg)
                    }
                }
            })
            app.mount('#container')
```

```
            }
        </script>
</head>

<body>
    <div id="container">
        <hello></hello>
        <hr>                    组件可以多次重复使用
        <hello></hello>
    </div>
</body>
</html>
```

组件在使用的时候可以类似于一个标签,而且可以重复使用。组件的内部包括多个部分,template 是模板结构,methods 是该组件的方法,data 是该组件的数据。该案例中点击 h2 标签会出现警示框,效果如图 7-1 所示。

图 7-1　Vue 3.0 组件

7.1.2　创建 Vue 2.0 组件

Vue 2.0 中组件的注册方法与 Vue 3.0 中不同,下面案例是使用 Vue 2.0 创建组件,具体代码如下:

```
<!DOCTYPE html>
<html lang="en">
<head>
    <meta charset="UTF-8">
    <title>Document</title>
    <script type="text/javascript" src="js/vue.js"></script>
    <script type="text/javascript">
        window.onload = function () {            Vue 2.0 中组件的注册方法

            Vue.component('mycom', {                    模板部分有两个标签,
                template: '<h1>组件</h1><h2>{{msg}}</h2>',   分别是 h1 和 h2
```

```
            data() {
                return {
                    msg: "vue2.0创建的组件"
                }
            }
        })
        new Vue({
            el: '#container',
        })

    }
    </script>
</head>
<body>
    <div id="container">
        <mycom></mycom>
    </div>
</body>
</html>
```

在案例中，模板部分有两个标签，分别是 h1 和 h2，但是案例执行的效果中只有 h1 标签，没有 h2 标签，如图 7-2 所示。在控制台中可以看到代码报错，如图 7-3 所示。报错的原因是在 Vue 2.0 中模板部分必须要有一个根节点。在 Vue 3.0 中模板中没有根节点，虽然没有报错，如图 7-4 所示，但还是推荐要在模板中添加一个根节点。

当模板部分的内容比较复杂时，推荐使用模板字符串，模板字符串是一对反引号（`），在模板字符串中代码可以任意换行，如果不使用模板字符串换行会报错。

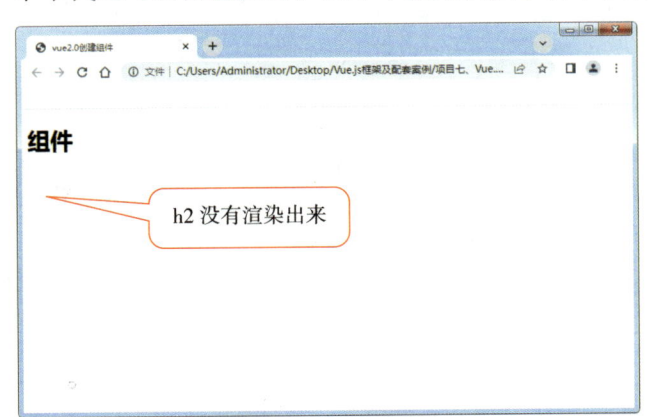

图 7-2　Vue 2.0 创建组件

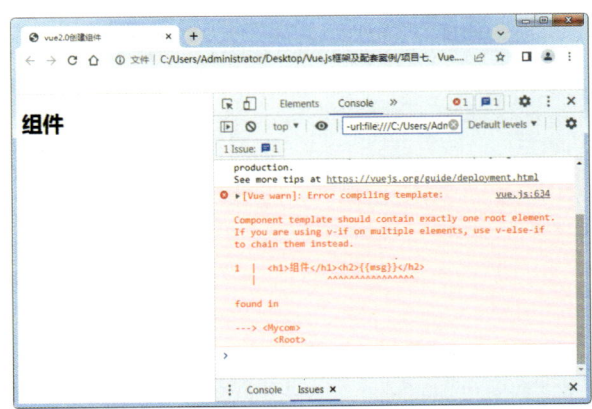

图 7-3　Vue 2.0 报错

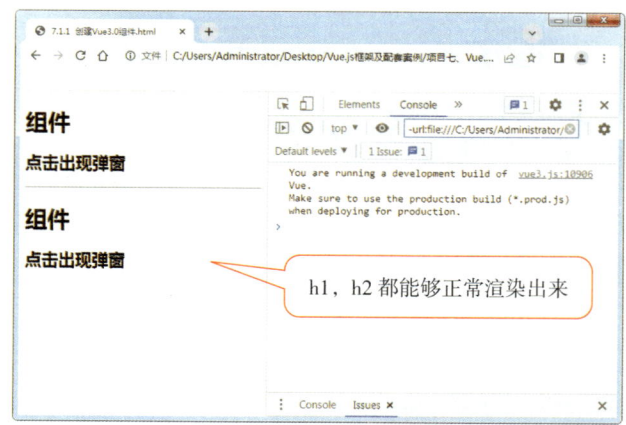

图 7-4　Vue 3.0 没有报错

模板部分修改以后的代码如下：

```
<script type="text/javascript">
        window.onload = function () {
            Vue.component('mycom', {
                template: `
                    <div>
                        <h1>组件</h1><h2>{{msg}}</h2>
                    </div>
                `,
                data() {
                    return {
                        msg: "vue2.0创建的组件"
                    }
                }
            })
            new Vue({
                el: '#container',
            })

        }
</script>
```

模板字符串（` `）

7.1.3　组件的命名

组件的命名：如果是小写字母，在使用的时候用原名称即可；如果除了首字母之外其他字母中有大写字母，在使用的时候要将名称进行一定的转换。下面案例定义了一个组件 sayHello，其代码如下：

```
<!DOCTYPE html>
<html lang="en">
<head>
```

```html
<meta charset="UTF-8">
<meta http-equiv="X-UA-Compatible" content="IE=edge">
<meta name="viewport" content="width=device-width, initial-scale=1.0">
<title>Document</title>
<script src="./js/vue3.js"></script>
<script>
    window.onload = function () {
        // 创建一个 Vue 应用
        const app = Vue.createApp({})
        app.component('sayHello', {        // 注册组件：sayHello
            template:
                `<div class="Say">
                    <h1>组件的命名</h1>
                    <h2>sayHello</h2>
                </div>
                `
        })
        app.mount('#container')
    }
</script>
</head>

<body>
    <div id="container">
        <sayHello></sayHello>          <!-- 正确的写法应该是 <say-hello></say-hello> -->
    </div>
</body>
</html>
```

案例执行后，控制台报错，组件没有正常加载出来，效果如图 7-5 所示。除了首字母之外的其他大写字母都要转换为连字符并加上对应的小写字母。例如，上述组件名为"sayHello"，在应用组件的时候要使用 <say-hello></say-hello>。如果组件名称为"Sayhello"，则在应用组件的时候使用原名称 < Sayhello ></ Sayhello > 即可。

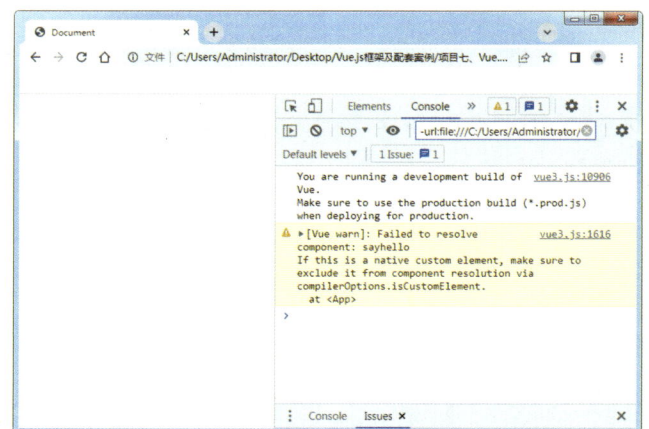

图 7-5 组件的命名

计数器组件

7.1.4 计数器组件

下面案例是一个计数器组件，通过该案例可以更加深刻地理解组件的复用性以及独立性。

设计思路：注册一个计数器组件，在组件中有一个变量 num，点击加号按钮可以增加 num 的值，最大值为 5；点击减号按钮可以减少 num 的值，最小值为 0。具体代码如下：

```
<!DOCTYPE html>
<html lang="en">
<head>
    <meta charset="UTF-8">
    <meta http-equiv="X-UA-Compatible" content="IE=edge">
    <meta name="viewport" content="width=device-width, initial-scale=1.0">
    <title>计数器组件</title>
    <style>
        .cnt .number{ padding: 0 10px }
    </style>
    <script src="./js/vue3.js"></script>
    <script>
        window.onload = function () {
            // 创建一个Vue 应用
            const app = Vue.createApp({})
            app.component('counter', {
                template:
                    `<div class="cnt">
                        <h2>计数器组件</h2>
                        <input type="button" value="-" @click="reduce">
                        <span class="number">{{num}}</span>
                        <input type="button" value="+" @click="add">
                    </div>
```

```
                `,
                data() {
                    return {
                        num:0
                    }
                },
                methods: {
                    add:function(){
                        if(this.num<10){
                            this.num++
                        }


                    },
                    reduce:function(){
                        if(this.num>0){
                            this.num--
                        }


                    }
                }
            })
            app.mount('#container')
        }
    </script>
</head>
<body>
    <div id="container">
        <counter></counter>
        <hr>
        <counter></counter>
    </div>
</body>
</html>
```
两个计数器组件，中间使用水平线间隔

在页面会出现两个计数器，每一个计数器相互独立，如图7-6所示。设想如果不使用组件，将会出现大量重复的代码，同时也会增加修改和维护的成本。

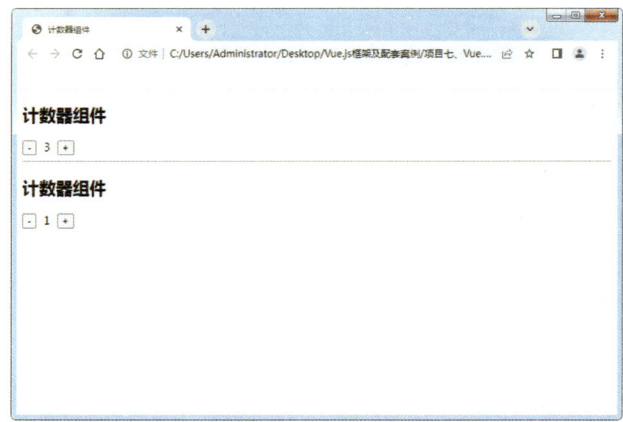

图 7-6 计数器组件

7.2 组件的切换

组件的切换可以使用 v-if 指令，也可以使用 Vue 提供的 component 标签，component 是一个占位符，使用 :is 属性来指定要展示的组件的名称。下面案例是登录方式的两个组件的切换。可以先将功能在主 Vue 对象中完成，然后再将代码封装到组件中，以降低组件开发的难度。

7.2.1 登录效果静态界面

在登录界面中有两个选项：账号登录和扫码登录。先在 HTML 和 CSS 中完成静态界面，其代码如下：

```
<!DOCTYPE html>
<html lang="en">
<head>
    <meta charset="UTF-8">
    <meta http-equiv="X-UA-Compatible" content="IE=edge">
    <meta name="viewport" content="width=device-width, initial-scale=1.0">
    <title>登录效果静态界面</title>
    <link rel="stylesheet" href="./css/style.css">
</head>
<body>
    <div id="container">
        <div class="login">
            <div class="login_nav">
                <ul>
                    <li class="on">账号登录</li>
                    <li>扫码登录</li>
                </ul>
            </div>
            <div class="login_cont">
```

```html
                <div class="way1">
                    <p><input type="text" placeholder="账号" class="txt"></p>
                    <p><input type="password" placeholder="密码" class="txt"></p>
                    <p><input type="button" value="登录" class="btn"></p>
                </div>
                <div class="way2">
                    <img src="./images/scane.jpg" alt="">
                </div>
            </div>
        </div>
    </div>
</body>
</html>
```

CSS 代码如下：

```css
*{margin: 0; padding: 0;}
.login{ width:400px; height:300px; margin: 50px auto; background-color: #333; padding: 10px;}
.login .login_nav{ height: 50px; margin-bottom: 20px;}
.login .login_nav li{ list-style: none; width: 200px; line-height:50px; float: left; text-align: center;
    color: #FFFFFF; border-bottom: 2px transparent solid; cursor: pointer; }
.login .login_nav li.on{ border-bottom: 2px #CC3366 solid;}
.login .login_cont .way1 p{ line-height: 50px;text-align: center;}
.login .login_cont .way1 .txt{ width: 90%; border-radius: 5px; height: 30px; border:1px #CCCCCC solid;padding-left: 10px; }
.login .login_cont .way1 .btn{ border: none; background-color: #CC3366; width: 50%; border: none; color: #FFFFFF; height: 30px; border-radius: 5px;}
.login .login_cont .way2{ text-align: center;}
```

两个登录方式在初始状态下都是显示状态，代码效果如图 7-7 所示。

图 7-7　登录效果静态界面

7.2.2 封装组件并进行切换

将两种登录方式分别封装成两个组件，设置变量com用来保存组件的名称，使用component标签的 :is 属性来指定要展示的组件的名称。点击"账号登录"或者"扫码登录"按钮改变com的值，从而进行组件的切换，具体代码如下：

```html
<!DOCTYPE html>
<html lang="en">
<head>
    <meta charset="UTF-8">
    <meta http-equiv="X-UA-Compatible" content="IE=edge">
    <meta name="viewport" content="width=device-width, initial-scale=1.0">
    <title>Document</title>
    <link rel="stylesheet" href="./css/style.css">
    <script src="./js/vue3.js"></script>
    <script>
        window.onload=function(){
            const app=Vue.createApp({
                data(){
                    return {
                        com:'way1'     // 设置变量com，记录当前组件名称
                    }
                }
            })
            app.component('way1',{       // 账号登录组件
                template:`
                <div class="way1">
                    <p><input type="text" placeholder="账号" class="txt"></p>
                    <p><input type="password" placeholder="密码" class="txt"></p>
                    <p><input type="button" value="登录" class="btn"></p>
                </div>
                `
            })
            app.component('way2',{       // 扫码登录组件
                template:`
                <div class="way2">
                    <img src="./images/scane.jpg" alt="">
                </div>
                `
            })
            app.mount('#container')
```

```html
            }
        </script>
    </head>
    <body>
        <div id="container">
            <div class="login">
                <div class="login_nav">
                    <ul>
                        <li :class="{on:com=='way1'}" @click="com='way1'">账号登录</li>
                        <li :class="{on:com=='way2'}" @click="com='way2'">扫码登录</li>
                    </ul>
                </div>
                <div class="login_cont">
                    <component :is="com"></component>
                </div>
            </div>
        </div>
    </body>
</html>
```

> :is 属性指定要展示的组件的名称

页面加载后的效果如图 7-8 所示。点击"扫码登录"后的效果如图 7-9 所示。

图 7-8　账号登录

图 7-9　扫码登录

7.2.3　选项卡组件切换

选项卡组件切换的设计思路与上面的登录组件切换相似。选项卡有三个选项，分别是头条、社会、娱乐。三个部分对应三个组件，这里对选项卡又进行了进一步的封装，封装为选项卡组件。其代码如下：

```html
<!DOCTYPE html>
<html lang="en">
<head>
    <meta charset="UTF-8">
```

```html
<meta http-equiv="X-UA-Compatible" content="IE=edge">
<meta name="viewport" content="width=device-width, initial-scale=1.0">
<title>Document</title>
<style>
    *{margin: 0; padding: 0;}
    body{ background-color: #CCCCCC;}
    .tab{ width:450px; height:203px; margin: 50px auto; background-color: rgb(255, 255, 255); overflow: hidden; }
    .tab .list{ height: 40px; background-image: linear-gradient(to bottom, #dbc755,#FFFFFF); margin-bottom: 3px;}
    .tab .list li{ list-style: none; width: 150px; line-height:40px; float: left; text-align: center;
    color: #000000; cursor: pointer; }
    .tab .list li.on{ border-bottom: 3px orange solid;}
    .toutiao,.shehui,.yule{ height: 170px; padding: 10px;}
    .toutiao{ background-color: lightblue;}
    .shehui{ background-color:lightgreen;}
    .yule{ background-color:rgb(241, 204, 210)}
</style>
<script src="./js/vue3.js"></script>
<script>
    window.onload=function(){
        const app=Vue.createApp({})
        app.component('toutiao',{         // toutiao 组件
            template:`
            <div class="toutiao">
                <h1> 头条 </h1>
                <p> 头条部分的内容 </p>
            </div>
            `
        })

        app.component('shehui',{          // shehui 组件
            template:`
            <div class="shehui">
                <h1> 社会 </h1>
                <p> 社会部分的内容 </p>
            </div>
            `
        })
```

```
        app.component('yule',{              ← yule 组件
            template:`
            <div class="yule">
                <h1>娱乐</h1>
                <p>娱乐部分的内容</p>
            </div>
            `
        })
        app.component('tab',{                ← tab 组件
            template:`
            <div class="tab">
                <div class="list">
                    <ul>
                        <li v-for="(item,index) in tabList"
                        @click="i=index" :class="{on:i==index}">
                            {{item.title}}
                        </li>
                    </ul>
                </div>
                <div class="tab_cont">
                    <component :is="tabList[i].com"></component>
                </div>
            </div>
            `,
            data(){
                return {                     ← 变量 tabList 设置了三个
                    tabList:[                   栏目与组件的对应关系
                        {title:"头条",com:"toutiao"},
                        {title:"社会",com:"shehui"},
                        {title:"娱乐",com:"yule"},
                    ],
                    i:0
                }
            }
        })

        app.mount('#container')
    }
    </script>
</head>
```

```
<body>
    <div id="container">
        <tab></tab>
<tab></tab>
    </div>
</body>
</html>
```

封装好的选项卡组件在应用时只要使用 <tab></tab> 标签即可,这里放置了两个选项卡组件,切换效果如图 7-10 所示。能否依然使用 <tab></tab> 组件实现在第二个选项卡组件中设置其他类型的栏目,如军事、新闻、影视等?要求的最终效果如图 7-11 所示。要实现图 7-11 的效果需要将选项和组件对应的变量放置在主 Vue 的 data 中,选项卡组件从主 Vue 组件中获取数据,这就要使用到组件的通信。

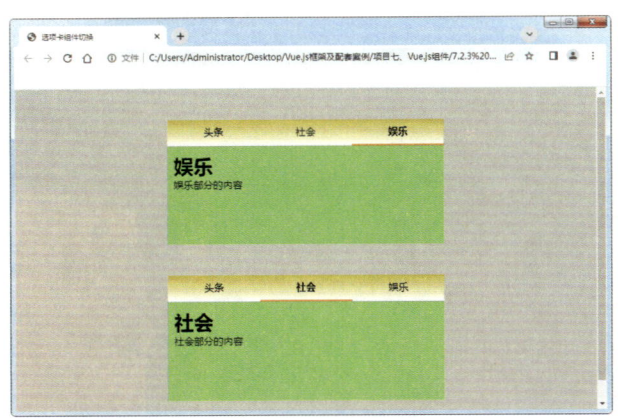

图 7-10 选项卡组件 1

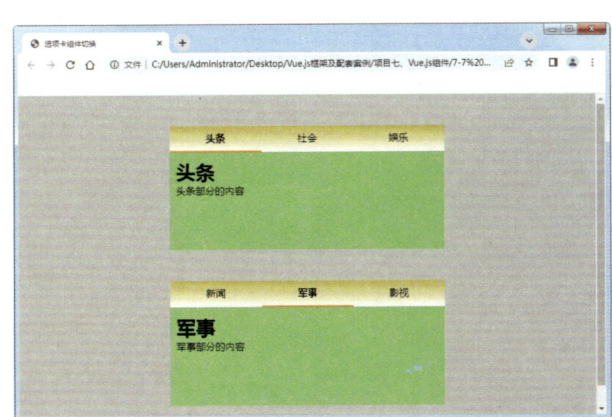

图 7-11 选项卡组件 2

7.3 组件的通信

组件的通信是指在组件之间传递数据的操作。组件的通信包括父组件向子组件传递数据,子组件向父组件传递数据,兄弟组件之间传递数据等多种类型。这里主要介绍父组件向子组件传递数据和子组件向父组件传递数据的方法。

7.3.1 父组件向子组件传递数据

父组件可以通过 props 向子组件传递数据。props 可以使用数组的形式在子组件中进行定义,props 中的元素表示父组件可以向子组件传递的数据项。在子组件中,可以将 props 中的元素理解为一个占位用的数据,在应用时从父组件中获得真正的数据。下面案例定义了两个组件,分别为 son 组件和 father 组件,father 组件内部包含了 son 组件,所以 son 组件是 father 组件的子组件,其代码如下:

```
<!DOCTYPE html>
<html lang="en">
<head>
    <meta charset="UTF-8">
    <meta http-equiv="X-UA-Compatible" content="IE=edge">
```

```html
<meta name="viewport" content="width=device-width, initial-scale=1.0">
<title>父组件向子组件传递数据</title>
<style>
    .father{ width: 700px; height: 400px; background-color: lightblue; padding: 30px;}
    .son{ width: 500px; height: 200px; background-color: lightpink;}
</style>
<script src="./js/vue3.js"></script>
<script>
    window.onload = function () {
        const app = Vue.createApp({})
        app.component("son", {          // son 组件
            template:`
                <div class="son">
                    <h2>子组件</h2>
                    <h3>父组件向子组件传递数据——hello,{{name}}</h3>
                </div>
            `,
            // 这里的 name 可以理解为一个占位用的假数据
            data() {
                return {
                }
            },
            props:["name"]   // props 中定义 name 元素，name 将会从父组件得到传递进来数据
        })

        app.component("father",{
            template:`
            <div class="father">
                <h1>父组件</h1>
                <p>父组件中有一个变量 fatherName,值是：小明</p>
                <son :name="fatherName"></son>
            </div>
            `,
            // 父组件将自己 data 中的 fatherName 变量传递给子组件的 name 元素
            data(){
                return {
                    fatherName:"小明"
                }
            }
```

```
            })
            app.mount('#container')
        }
    </script>
</head>
<body>
    <div id="container">
        <father></father>
    </div>
</body>
</html>
```

父组件向子组件传递数据，即表示子组件使用父组件中的数据，可以分为以下几个步骤：

（1）在子组件中使用 props 定义元素，如案例中的 name；

（2）子组件使用 name 来完成特定功能，但是要理解这里的 name 只相当于一个占位用的假数据；

（3）在父组件内部的那个子组件中使用 :name= 父组件中的变量，即 name 得到父组件中的某个变量，完成父组件向子组件传递数据的任务。

7.3.2 完成选项卡组件

为了提高选项卡组件的通用性，将之前的选项卡组件案例进行修改。在主 Vue 对象中定义多个选项卡对应数据 tablist1 和 tablist2，在 tab 组件中使用 props 定义 tablist，然后由父组件向 tab 组件传递数据，具体代码如下：

```
<!DOCTYPE html>
<html lang="en">

<head>
    <meta charset="UTF-8">
    <meta http-equiv="X-UA-Compatible" content="IE=edge">
    <meta name="viewport" content="width=device-width, initial-scale=1.0">
    <title>选项卡组件切换</title>
    <style>
        *{margin: 0; padding: 0;}
        body{ background-color: #CCCCCC;}
        .tab{ width:450px; height:203px; margin: 50px auto; background-color: rgb(255, 255, 255); overflow: hidden; }
        .tab .list{ height: 40px; background-image: linear-gradient(to bottom,#dbc755,#FFFFFF); margin-bottom: 3px;}
        .tab .list li{ list-style: none; width: 150px; line-height:40px; float: left; text-align: center;
        color: #000000;  cursor: pointer; }
```

```
        .tab .list li.on{ border-bottom: 3px orange solid; font-weight: bold;}
        .tab .tab_cont{height: 170px; padding: 10px; background-color:lightgreen;}
</style>
<script src="./js/vue3.js"></script>
<script>
    window.onload=function(){
        const app=Vue.createApp({
            data(){
                return {
                    tablist1:[
                        {title:"头条",com:"toutiao"},
                        {title:"社会",com:"shehui"},
                        {title:"娱乐",com:"yule"},
                    ],
                    tablist2:[
                        {title:"新闻",com:"xinwen"},
                        {title:"军事",com:"junshi"},
                        {title:"影视",com:"yingshi"},
                    ],
                }
            }
        })

        app.component('toutiao',{
            template:`
            <div class="toutiao">
                <h1> 头条 </h1>
                <p> 头条部分的内容 </p>
            </div>
            `
        })
         app.component('shehui',{
            template:`
            <div class="shehui">
                <h1> 社会 </h1>
                <p> 社会部分的内容 </p>
            </div>
            `
        })
```

> Vue 根对象是最大的根组件，在 Vue 根对象中定义变量 tablist1 和 tablist2

```
    template:`
    <div class="yule">
        <h1>娱乐</h1>
        <p>娱乐部分的内容</p>
    </div>
    `
})
app.component('xinwen',{
    template:`
    <div class="xinwen">
        <h1>新闻</h1>
        <p>新闻部分的内容</p>
    </div>
    `
})
app.component('junshi',{
    template:`
    <div class="junshi">
        <h1>军事</h1>
        <p>军事部分的内容</p>
    </div>
    `
})
app.component('yingshi',{
    template:`
    <div class="yingshi">
        <h1>影视</h1>
        <p>影视部分的内容</p>
    </div>
    `
})
app.component('tab',{
    template:`
    <div class="tab">
        <div class="list">
            <ul>
                <li  v-for="(value,index) in tablist" :class="{on:i==index}"
                    @click="i=index">
                    {{value.title}}
                </li>
```

```
                </ul>
            </div>
            <div class="tab_cont">
                <component :is="tablist[i].com"></component>
            </div>
        </div>
    `,
    data(){
        return {
            i:0,
        }
    },
    methods:{},
    props:['tablist']   ← 在 tab 组件中使用 props
})
app.mount('#container')
    }
    </script>
</head>
<body>
    <div id="container">
        <tab :tablist="tablist1"></tab>   ← 父组件向子组件传递数据
        <tab :tablist="tablist2"></tab>
    </div>
</body>
</html>
```

虽然只定义了一个 tab 组件，但是当传递过来的数据不同，tab 组件显示的内容就不同，这充分体现了组件的复用性及灵活性。同时 props 是单向数据流的，它只能从父组件传递到子组件，而子组件是无法更改 props 的值的，只能由父组件来修改。这样就保证了组件的数据传递不会出现混乱的情况。案例效果如图 7-12 所示。

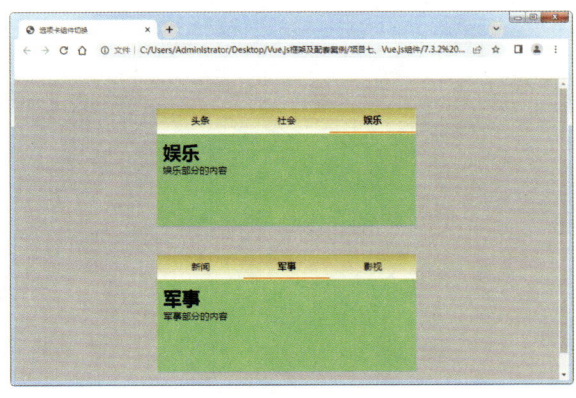

图 7-12　完成选项卡组件

7.3.3　子组件向父组件传递数值的方法——ref 方法

子组件向父组件传递数值有两种方法，一种是使用 ref，其特点是简单直接；另一种是使用自定义事件，虽然步骤较多，但是更为通用。

ref 可以给 DOM 节点添加也可以给组件添加，可以理解为使用 ref 起一个名字，然后通过 Vue 实例的 $refs 属性找到对应名字的 DOM 节点或者组件。下面案例中有一个父组件 father 和一个子组件 son，点击按钮可以从子组件向父组件中传递数值。具体代码如下：

```html
<!DOCTYPE html>
<html lang="en">
<head>
    <meta charset="UTF-8">
    <meta http-equiv="X-UA-Compatible" content="IE=edge">
    <meta name="viewport" content="width=device-width, initial-scale=1.0">
    <title>父组件使用子组件中的值</title>
    <style>
        .father{ width: 700px; height: 400px; background-color: lightblue; padding: 30px;}
        .son{ width: 500px; height: 200px; background-color: lightpink;}
    </style>
    <script src="./js/vue3.js"></script>
    <script>
        window.onload = function () {
            // 创建一个 Vue 应用
            const app = Vue.createApp({})
            app.component("son", {
                template:`
                    <div class="son">
                        <h2>子组件</h2>
                        <p>子组件中有 2 个变量 sonName 的值是：小红, age 的值是 :10</p>
                    </div>
                `,
                data() {
                    return {
                        sonName:"小红",    // son 组件中的变量 :sonName
                        sonAge:10
                    }
                },
            })

            app.component("father",{
```

```
            template:`
            <div class="father">
                <h1>父组件</h1>
                <son ref="son1"></son>
                <input type="button" value="获取子组件的值" @click="getData()">
                <p>{{info}}</p>
            </div>
            `,
            data(){
                return {
                    info:""
                }
            },
            methods:{
                getData(){
                    this.info=this.$refs.son1.sonName
                }
            },
        })
        app.mount('#container')
        }
    </script>
</head>
<body>
    <div id="container">
        <father></father>
    </div>
</body>
</html>
```

（对 son 组件使用 ref 进行命名）

（使用 $refs 找到对应的 son 组件）

页面加载后的效果如图 7-13 所示。点击"获取子组件的值"按钮后，子组件向父组件传递 sonName 的值，其效果如图 7-14 所示。

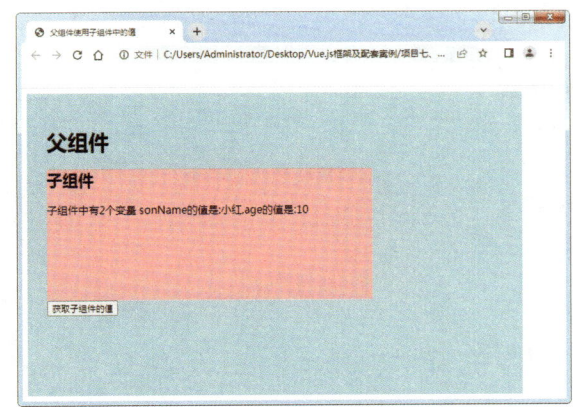

图 7-13　初始状态

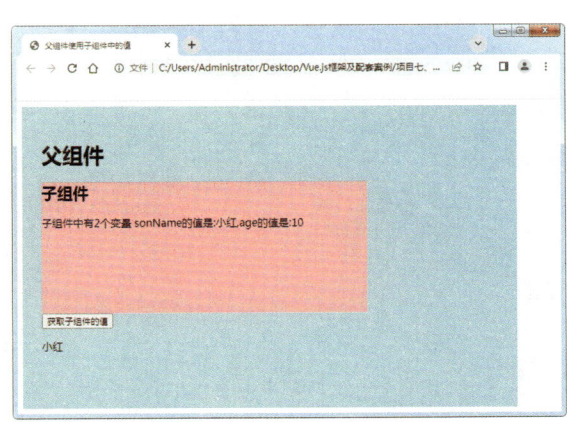

图 7-14　子组件向父组件传值

7.3.4 子组件向父组件传递数值的方法——自定义事件方法

子组件向父组件传递数值通过 $emit 来实现。$emit 绑定一个自定义事件 event，父组件通过 @event 监听并接收事件传递过来的参数。下面案例中子组件 son 中点击按钮会触发 send 函数，在 send 函数中使用 $emit 绑定自定义事件 func，并传递参数。在父组件中应用 son 组件，在 son 组件上添加 func 自定义事件，父组件会监听 func 事件，当 func 事件触发的时候再执行父组件中的 getData() 函数。具体代码如下：

```
<!DOCTYPE html>
<html lang="en">
<head>
    <meta charset="UTF-8">
    <meta http-equiv="X-UA-Compatible" content="IE=edge">
    <meta name="viewport" content="width=device-width, initial-scale=1.0">
    <title> 父组件使用子组件中的值 </title>
    <style>
        .father{ width: 700px; height: 400px; background-color: lightblue; padding: 30px;}
        .son{ width: 500px; height: 200px; background-color: lightpink;}
    </style>
    <script src="./js/vue3.js"></script>
    <script>
        window.onload = function () {
            // 创建一个 Vue 应用
            const app = Vue.createApp({})
            app.component("son", {
                template:`
                    <div class="son">
                        <h2> 子组件 </h2>
                        <p> 子组件中有 2 个变量 sname 的值是：小红 ,age 的值是 :10</p>
                        <input type="button" value=" 将数据传递给父组件 "
                        @click="send()">
                    </div>
                `,
                data() {
                    return {
                        sonName:" 小红 ",
                        sonAge:10
                    }
                },
                methods:{
                    send(){
```

> 点击 son 组件的按钮触发 send() 函数

```
                    this.$emit("func",{
                        name:this.sonName,
                        age:this.sonAge
                    })
                }
            }
        })
        app.component("father",{
            template:`
            <div class="father">
                <h1>父组件</h1>
                <son @func="getData($event)"></son>
                从子组件中传递过来的值是：
                <p>{{info.name}}</p>
                <p>{{info.age}}</p>
            </div>
            `,
            data(){
                return {
                    info:""
                }
            },
            methods:{
                getData(value){
                    this.info=value
                }
            },
        })
        app.mount('#container')
    }
    </script>
</head>
<body>
    <div id="container">
        <father></father>
    </div>
</body>
</html>
```

> 在 send() 函数中使用 $emit 绑定自定义事件 func，并传递参数

> 当 func 事件被触发时调用 getData() 函数，$event 是传递进来的参数

> value 是 $event 传递进来参数，将其赋值给父组件的 info

@func 是自定义事件，可以类比于 @click、@mouseover 等来理解。页面加载以后的效果如图 7-15

所示。点击按钮以后，子组件通过参数的形式向父组件传递数值，其效果如图 7-16 所示。

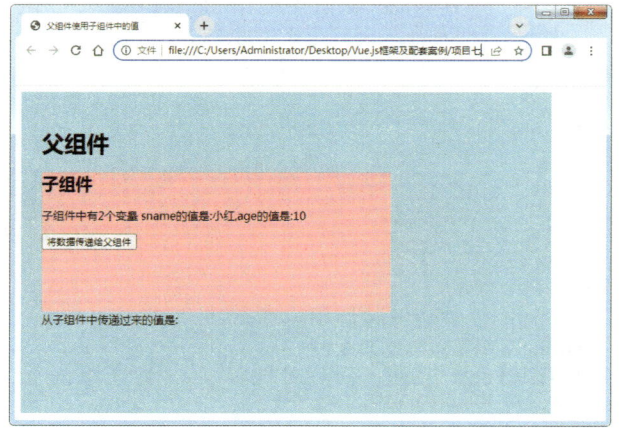

图 7-15 初始状态

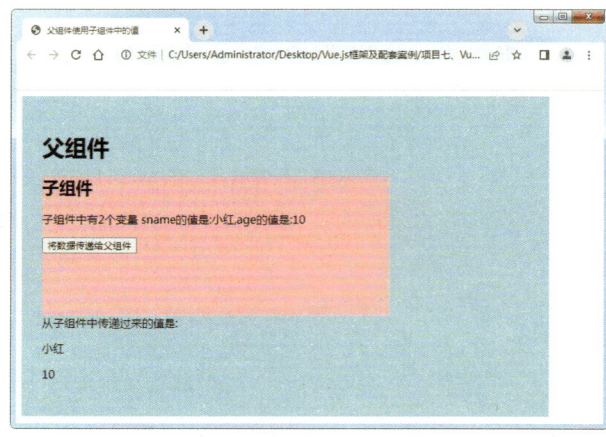

图 7-16 子组件向父组件中传递数值

使用自定义事件传递参数可以有两种方法：第一种是使用上面案例中的 $event，这里的 $event 不能换成其他名称的参数；第二种是不写参数，但是这种写法的函数必须不能带有括号。father 组件的具体代码如下：

```
app.component("father",{
    template:`
    <div class="father">
        <h1>父组件</h1>
        <son @func="getData"></son>
        从子组件中传递过来的值是：
        <p>{{info.name}}</p>
        <p>{{info.age}}</p>

    </div>
    `,
    data(){
        return {
            fname:"小明",
            info:""
        }
    },
    methods:{
        getData(value){
            this.info=value
        }
    },
})
```

> getData 没有参数也不能在后面带上括号。这种方法会传递默认参数，即的第一个参数值

> 使用 value 接收默认参数

7.4 插 槽

7.4.1 默认插槽

插槽是组件封装期间预留的内容占位符，在使用该组件时可以为插槽指定填充的内容。插槽使用 slot 来定义。插槽分为多种类型，下面分别介绍默认插槽、具名插槽和作用域插槽。

默认插槽的使用方法是将填充的内容放置在组件标签之间。下面案例中定义了一个 error 组件，可以通过 slot 指定不同类型的错误信息，具体代码如下：

```
<!DOCTYPE html>
<html lang="en">
<head>
    <meta charset="UTF-8">
    <meta http-equiv="X-UA-Compatible" content="IE=edge">
    <meta name="viewport" content="width=device-width, initial-scale=1.0">
    <title>Document</title>
    <style>
        .error{ width: 400px; height: 100px; background-color: lightblue; border-radius: 5px; text-align: center; margin: 20px;}
    </style>
    <script src="./js/vue3.js"></script>
    <script>
        window.onload = function () {
            const app = Vue.createApp({})
            app.component('error', {
                template:`
                <div class="error">
                    <h2>ERROR!</h2>
                    <slot>默认错误</slot>    ← slot 部分是可以被替换的
                </div>
                `,
            })
            app.mount('#container')
        }
    </script>
</head>

<body>
    <div id="container">
```

```
            <error></error>
            <error>缺少参数！</error>
            <error>参数格式错误！</error>
        </div>
    </body>
</html>
```

<error></error> 内部没有内容时会显示 <slot></slot> 中的内容，<error></error> 内部有内容时，内部的内容会替换 <slot></slot> 中的内容。上述代码执行后的效果如图7-17所示。

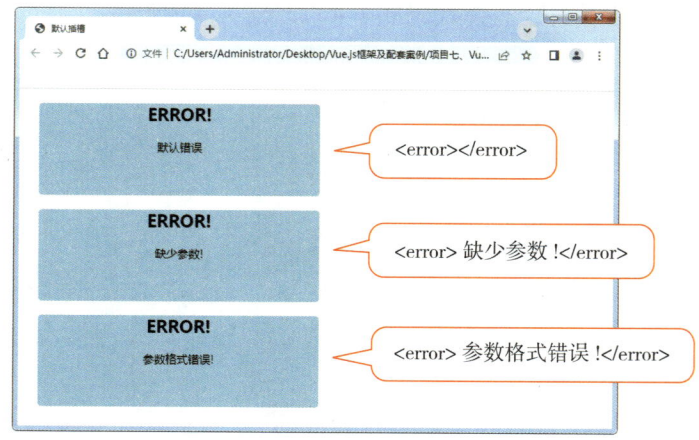

图7-17　默认插槽

7.4.2　具名插槽

具名插槽，顾名思义就是有名字的插槽，可以通过名字来进行插槽的调用。插槽使用 name 来进行名称的定义，使用时通过 <template v-slot: 插槽名 ></template > 来进行调用。下面案例中定义了 layout 组件，其内部有三个具名插槽，分别是 header、main、footer，具体代码如下：

```
<!DOCTYPE html>
<html lang="en">
<head>
    <meta charset="UTF-8">
    <meta http-equiv="X-UA-Compatible" content="IE=edge">
    <meta name="viewport" content="width=device-width, initial-scale=1.0">
    <title>具名插槽</title>
    <style>
        .layout{ width: 600px; height: 200px; background-color:lightgreen; border-
        radius: 5px; text-align: center; margin: 20px;}
    </style>
    <script src="./js/vue3.js"></script>
    <script>
        window.onload = function () {
            const app = Vue.createApp({})
```

```
            app.component('layout', {
                template:`
                <div class="layout">
                    <div>
                        <slot name="header"></slot>
                    </div>
                    <div>
                        <slot name="main"></slot>       ← 三个具名插槽
                    </div>
                    <div>
                        <slot name="footer"></slot>
                    </div>
                </div>
                `,
            })
            app.mount('#container')
        }
    </script>
</head>
<body>
    <div id="container">
        <layout>
            <template v-slot:header>        ← header 插槽的使用
                <h3>layout 的头部 </h4>
            </template>
            <template v-slot:main>
                <p>layout 的内容 </p>
            </template>
            <template v-slot:footer>
                <h4>layout 的尾部 </h4>
            </template>
        </layout>

        <layout>
            <template v-slot:footer>
                <h4>layout 的尾部 </h4>       ← 这个 layout 组件中几个具
            </template>                          名插槽的顺序是打乱的
            <template v-slot:header>
                <h3>layout 的头部 </h4>
            </template>
```

```
            <template v-slot:main>
                <p>layout 的内容 </p>
            </template>
        </layout>
    </div>
</body>
</html>
```

在案例中有两个 layout 组件，第二个 layout 组件在应用的时候将几个具名插槽的顺序打乱，加载出来后的效果依然和组件中定义的顺序相同，这使得具名插槽在使用时能够更加灵活。案例代码执行后的效果如图 7-18 所示。

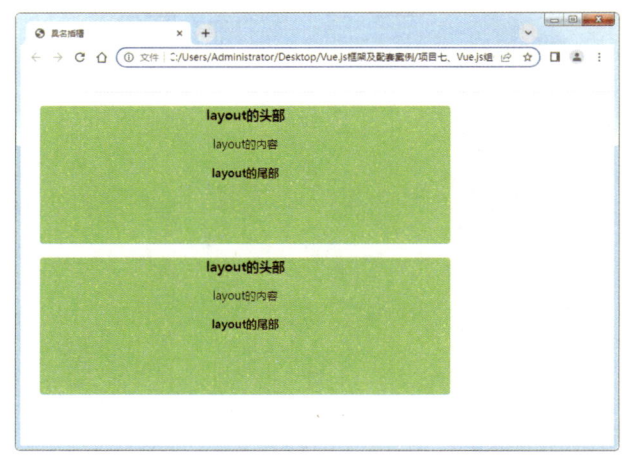

图 7-18　具名插槽

7.4.3　作用域插槽

当子组件中定义了数据，父组件要使用子组件的这些数据时，因为作用域的问题，父组件无法直接得到子组件的这些数据，而使用作用域插槽可以解决这个问题。

作用域插槽要先在子组件中使用插槽绑定自己 data 中的数据，在父组件应用的时候使用 v-slot: 插槽名称 =" 接收数据的变量 "。下面案例中定义了一个 student 组件，其中内部有两个变量 info 和 score，插槽的 name 是 "stu"，在插槽中使用 :stuInfo="info" 和 :stuScore="score" 分别来绑定两个变量。在应用时使用 v-slot:stu="scope" 来接收插槽中绑定的所有数据，即 scope 可以接收 stuInfo 和 stuScore 的值。scope 是一种常用的写法，也可以换成其他的名称。具体代码如下：

```
<!DOCTYPE html>
<html lang="en">
<head>
    <meta charset="UTF-8">
    <meta http-equiv="X-UA-Compatible" content="IE=edge">
    <meta name="viewport" content="width=device-width, initial-scale=1.0">
    <title>作用域插槽 </title>
    <style>
```

```
            .student{ width: 400px; height:100px; background-color:lightgreen; border-
            radius: 5px;margin:20px }
        </style>
        <script src="./js/vue3.js"></script>
        <script>
            window.onload = function () {
                const app = Vue.createApp({})
                app.component('student', {
                    template:`
                    <div class="student">
                            <slot name="stu" :stuInfo="info" :stuScore="score">
                            {{info.name}}</slot>
                    </div>
                    `,
                    data(){
                        return {
                            info:{
                                name:"小明",
                                sex:"male",
                                age:16,
                                class:"Class Two"
                            },
                            score:{
                                Chinese:96,
                                math:100,
                                English:99
                            }
                        }
                    }
                })
                app.mount('#container')
            }
        </script>
    </head>
    <body>
        <div id="container">
            <student></student>
            <student v-slot:stu="scope">年龄:{{scope.stuInfo.age}}</student>
            <student v-slot:stu="scope">数学成绩:{{scope.stuScore.math}}</student>
```

> 绑定 info 变量和 score 变量

> scope 接收绑定的全部数据

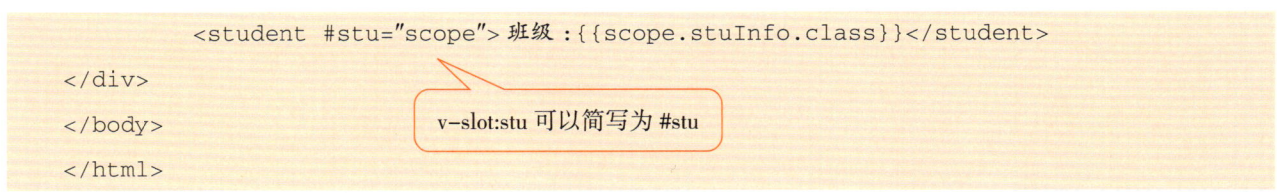

```
        <student #stu="scope">班级:{{scope.stuInfo.class}}</student>
    </div>
  </body>
</html>
```

> v-slot:stu 可以简写为 #stu

第一个 student 组件使用的是插槽中的默认值,显示的是学生的姓名。其他几个 student 组件使用作用域插槽分别显示了年龄、数学成绩和班级。该案例代码执行后的效果如图 7-19 所示。

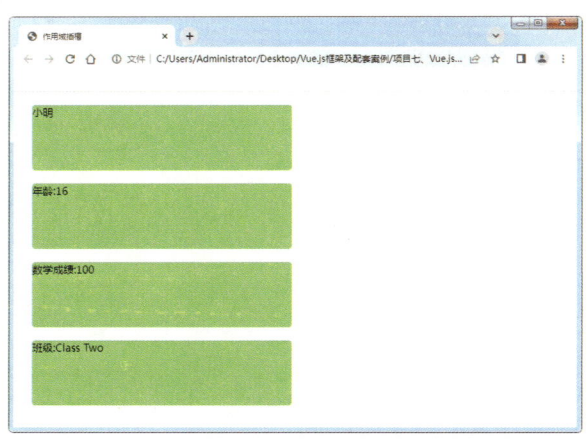

图 7-19　作用域插槽

7.5　综合实训：使用组件实现购物车功能

使用组件实现购物车功能项目中主要应用的知识点是组件的通信,其中既涉及子组件向父组件传递数据,也涉及父组件向子组件传递数据。

7.5.1　项目描述

使用组件实现购物车功能项目将购物车功能分为三个子组件,分别是用户名部分、产品数量部分和结算部分。改变产品数量时,总价会发生改变。点击删除按钮时,对应的产品会被删除。该项目最终效果如图 7-20 所示。

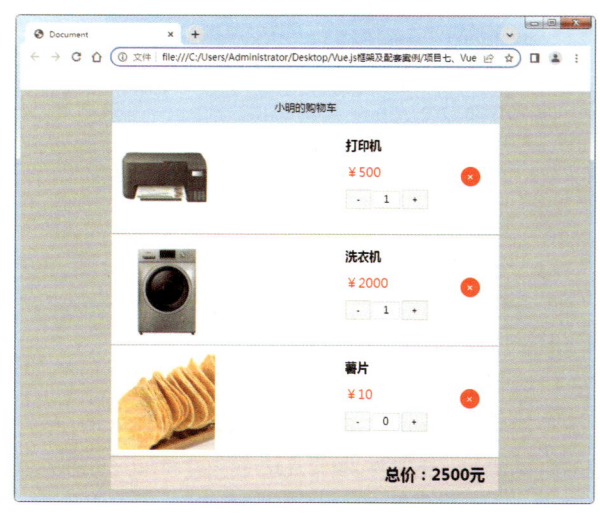

图 7-20　购物车效果

7.5.2 项目分析

父组件中放置各个产品的详细数据，子组件中要使用 props 来实现父组件向子组件传递数值的要求。点击"+"或者"-"按钮可以改变产品的数量，此事件是在子组件中发生的，要使用自定义事件的方法来实现子组件向父组件来传递数值的要求。

7.5.3 项目实施

项目实施步骤 1：完成购物车静态界面

购物车静态界面的代码如下：

```html
<!DOCTYPE html>
<html lang="en">
<head>
    <meta charset="UTF-8">
    <meta http-equiv="X-UA-Compatible" content="IE=edge">
    <meta name="viewport" content="width=device-width, initial-scale=1.0">
    <title> 购物车 </title>
    <link rel="stylesheet" href="./css/style.css">
</head>
<body>
    <div class="cont">
        <div class="shop">
            <div class="title"> 小明的购物车 </div>
            <div class="shop_cont">
                <ul>
                    <li>
                        <img src="./images/1.jpg" class="pic">
                        <div class="txt">
                            <h3> 产品标题 </h3>
                            <div class="jiage">
                                <p> ¥100</p>
                                <div class="num">
                                    <input type="button" value="-" class="btn">
                                    <span class="shu">0</span>
                                    <input type="button" value="+" class="btn">
                                </div>
                                <i class="shan">&times;</i>
                            </div>
                        </div>
                    </li>
```

```html
            <li>
                <img  src="./images/2.jpg"  class="tu">
                <div class="txt">
                    <h3>产品标题</h3>
                    <div class="jiage">
                        <p>¥100</p>
                        <div class="num">
                            <input type="button" value="-" class="btn">
                            <span class="shu">0</span>
                            <input type="button" value="+" class="btn">
                        </div>
                        <i class="shan">&times;</i>
                    </div>
                </div>
            </li>
            <li>
                <img  src="./images/3.jpg"  class="tu">
                <div class="txt">
                    <h3>产品标题</h3>
                    <div class="jiage">
                        <p>¥100</p>
                        <div class="num">
                            <input type="button" value="-" class="btn">
                            <span class="shu">0</span>
                            <input type="button" value="+" class="btn">
                        </div>
                        <i class="shan">&times;</i>
                    </div>
                </div>
            </li>
        </ul>
    </div>

    <div class="zongjia">
        <h2>总价：<span>0</span>元</h2>
    </div>
  </div>
</div>
</body>
</html>
```

CSS 代码如下：

```css
*{margin: 0; padding: 0;}
body{ background-color: #CCCCCC;}
.shop{ width: 600px; margin: 0 auto; background-color: #FFFFFF;}
.shop .title{ height: 50px; background-color: rgb(200, 223, 243); line-height: 50px; text-align: center;}
.shop_cont li{ height: 150px; padding: 10px; border-bottom: 1px #999 solid; list-style: none; position: relative;}
.shop_cont li .pic{ width: 150px; height: 150px; float: left ; margin-right: 15px;}
.shop_cont li .txt{ height: 150px; float:right; margin-right: 100px; }
.shop_cont li .txt h3{ margin-bottom: 15px; margin-top: 10px;}
.shop_cont li .txt .jiage p{ font-size: 20px; color: #F00;margin-bottom: 15px;}
.shop_cont li .txt .jiage .btn{ width: 40px; height: 30px; border: 1px #CCC solid;}
.shop_cont li .txt .shu{width: 40px; height: 25px; border: 1px #CCC solid; text-align: center; margin: 0 3px; display: inline-block;}
.shop_cont li .txt .shan{ position: absolute; right: 30px; top:65px; width: 30px; height: 30px; background-color: rgb(255, 0, 0); text-align: center; line-height: 30px;; text-decoration: none; border-radius: 50%; cursor: pointer; color: #FFF; cursor: pointer; font-style: normal;}
.shop .zongjia{ height: 50px; background-color: rgb(236, 222, 222); line-height: 50px; text-align: right; padding-right: 20px;  clear: both;}
```

购物车静态界面效果如图 7-21 所示。

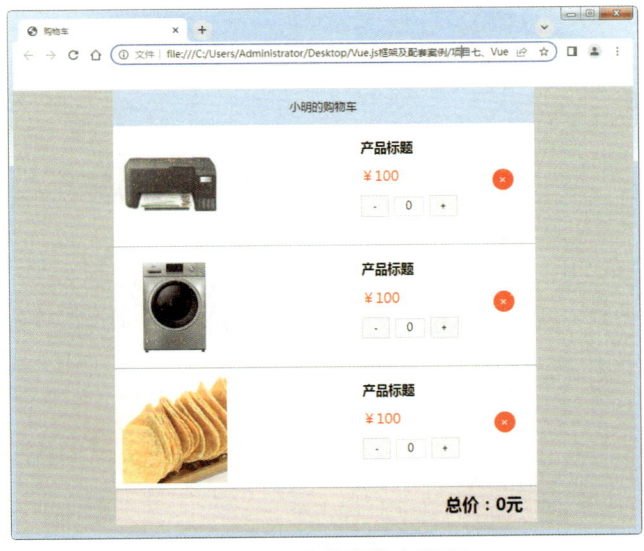

图 7-21 购物车静态界面

项目实施步骤 2：循环数据

项目实施步骤 2：循环数据

产品的详细数据放在主 Vue 对象中，使用 v-for 循环将数据渲染出来，同时使用 computed 来实现计算总价的功能。具体代码如下：

```html
<!DOCTYPE html>
<html lang="en">
<head>
    <meta charset="UTF-8">
    <meta http-equiv="X-UA-Compatible" content="IE=edge">
    <meta name="viewport" content="width=device-width, initial-scale=1.0">
    <title>购物车 步骤2</title>
    <link rel="stylesheet" href="./css/style.css">
    <script src="./js/vue3.js"></script>
    <script>
        window.onload = function () {
            const app = Vue.createApp({
                data() {
                    return {
                        uname: "小明",
                        list: [
                            {
                                id: 1,
                                name: '打印机',
                                price: 500,
                                num: 1,
                                src: './images/1.jpg'
                            },
                            {
                                id: 2,
                                name: '洗衣机',
                                price: 2000,
                                num: 1,
                                src: './images/2.jpg'
                            },
                            {
                                id: 3,
                                name: '薯片',
                                price: 10,
                                num: 0,
                                src: './images/3.jpg'
                            },
                        ]
                    }
```

各类产品的详细数据

```
                },
                computed:{
                    totalPrice(){
                        var t=0
                        this.list.forEach(item=>{
                            t=t+(item.price*item.num)
                        })
                        return t
                    }
                }
            })
            app.mount('#container')
        }
    </script>
</head>

<body>
    <div id="container">
        <div class="shop">
            <div class="title">{{uname}}的购物车</div>

            <div class="shop_cont">
                <ul>
                    <li v-for="item in list">
                        <img :src="item.src" class="pic">
                        <div class="txt">
                            <h3>{{item.name}}</h3>
                            <div class="jiage">
                                <p>￥{{item.price}}</p>
                                <div class="num">
                                    <input type="button" value="-" class="btn">
                                    <span class="shu">{{item.num}}</span>
                                    <input type="button" value="+" class="btn">
                                </div>
                                <i class="shan">&times;</i>
                            </div>
                        </div>
                    </li>
                </ul>
```

> 对 list 进行循环，求出产品总价格

> 循环 list，渲染数据

```
            </div>
            <div class="zongjia">
                <h2>总价：<span>{{totalPrice}}</span>元</h2>
            </div>
        </div>
    </div>
</body>
</html>
```

> 使用计算属性 totalPrice 来渲染出产品总价

上述代码执行后的效果如图 7-22 所示。

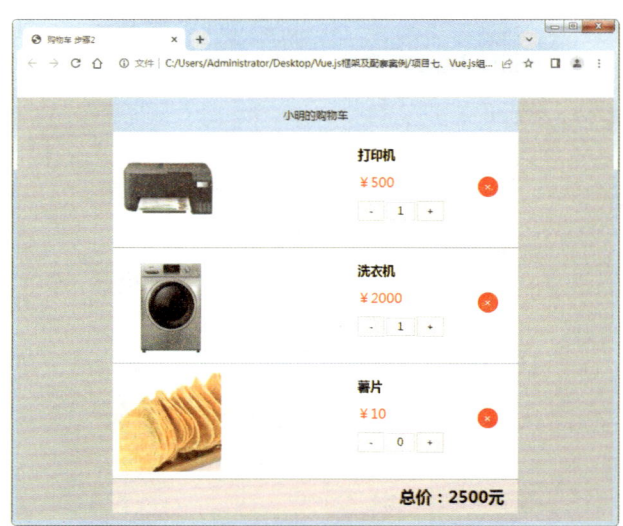

图 7-22　计算总价

项目实施步骤 3：封装组件

项目实施步骤 3：封装组件

步骤 2 中所有的数据和功能都是在主 Vue 对象中实现的，这样可以降低开发的难度，也方便调试错误。在步骤 3 中要将购物车的三部分分别封装成组件，数据依然放在主 Vue 对象中，即父组件中。子组件要使用 props 来完成父组件向子组件传递数据的任务。具体代码如下：

```
<!DOCTYPE html>
<html lang="en">
<head>
    <meta charset="UTF-8">
    <meta http-equiv="X-UA-Compatible" content="IE=edge">
    <meta name="viewport" content="width=device-width, initial-scale=1.0">
    <title>购物车　步骤 3</title>
    <link rel="stylesheet" href="./css/style.css">
    <script src="./js/vue3.js"></script>
    <script>
        window.onload = function () {
            const app = Vue.createApp({
```

```
        data() {
            return {
                uname:"小明",
                list: [
                    {
                        id: 1,
                        name: '打印机',
                        price: 500,
                        num: 1,
                        src: './images/1.jpg'
                    },
                    {
                        id: 2,
                        name: '洗衣机',
                        price: 2000,
                        num: 1,
                        src: './images/2.jpg'
                    },
                    {
                        id: 3,
                        name: '薯片',
                        price: 10,
                        num: 0,
                        src: './images/3.jpg'
                    },
                ]
            }
        }
})
app.component("shophead", {          // 购物车头部组件:shophead
    template: `
        <div class="title">{{uname}} 的购物车 </div>
    `,
    props:['uname']                  // 在 props 中定义 uname
})

    app.component("shopcont", {      // 购物车中部组件 : shopcont
        template: `
```

```
            <div class="shop_cont">
                <ul>
                    <li v-for="item in list">
                        <img :src="item.src" class="pic">
                        <div class="txt">
                            <h3>{{item.name}}</h3>
                            <div class="jiage">
                                <p>¥{{item.price}}</p>
                                <div class="num">
                                    <input type="button" value="-" class="btn">
                                    <span class="shu">{{item.num}}</span>
                                    <input type="button" value="+" class="btn">
                                </div>
                                <i class="shan">&times;</i>
                            </div>
                        </div>
                    </li>
                </ul>
            </div>
        `,
        props:['list']
    })


    app.component("shopfoot", {
        template: `
            <div class="zongjia">
                <h2>总价：<span>{{totalPrice}}</span>元</h2>
            </div>
        `,
        props:['list'],
        computed:{
            totalPrice(){
                var t=0
                this.list.forEach(item=>{
                    t=t+(item.price*item.num)
                })
                return t
```

```
                    }
                }
            })
            app.mount('#container')
        }
    </script>
</head>
<body>
    <div id="container">
        <div class="shop">
            <shophead :uname="uname"></shophead>
            <shopcont :list="list"></shopcont>
            <shopfoot :list="list"></shopfoot>
        </div>
    </div>
</body>
</html>
```

- props 中定义的 uname
- 主 Vue 对象 data 中的 uname 变量

购物车由三个组件组成，每一部分的数据都是从父组件传递给这三个子组件的。上述代码执行后的效果如图 7-23 所示。

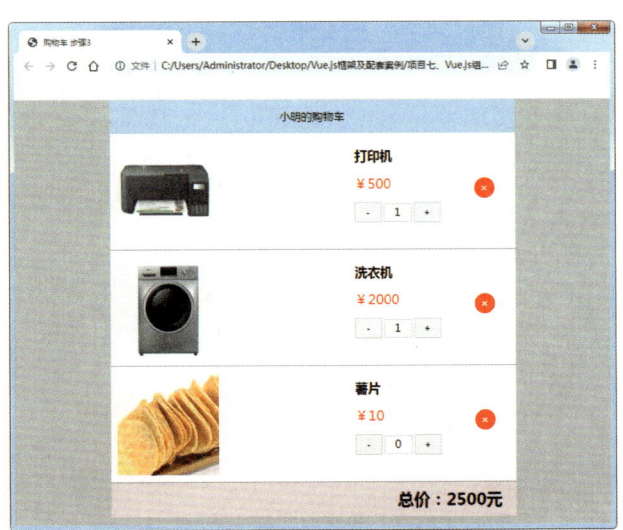

图 7-23　购物车封装成组件

项目实施步骤 4：添加自定义事件

在 shopcont 组件中，通过点击按钮可以改变产品的数量，子组件要向父组件传递新的产品数量，这就需要使用自定义事件来完成此功能。点击删除按钮能够将产品进行删除同样也需要使用自定义事件。其代码如下：

```html
<!DOCTYPE html>
<html lang="en">

<head>
    <meta charset="UTF-8">
    <meta http-equiv="X-UA-Compatible" content="IE=edge">
    <meta name="viewport" content="width=device-width, initial-scale=1.0">
    <title>购物车 步骤4</title>
    <link rel="stylesheet" href="./css/style.css">
    <script src="./js/vue3.js"></script>
    <script>
        window.onload = function () {
            const app = Vue.createApp({
                data() {
                    return {
                        uname: "小明",
                        list: [
                            {
                                id: 1,
                                name: '打印机',
                                price: 500,
                                num: 1,
                                src: './images/1.jpg'
                            },
                            {
                                id: 2,
                                name: '洗衣机',
                                price: 2000,
                                num: 1,
                                src: './images/2.jpg'
                            },
                            {
                                id: 3,
                                name: '薯片',
                                price: 10,
                                num: 0,
                                src: './images/3.jpg'
                            },
                        ]
```

```js
            }
        },
        methods:{
            numReduce(value){// 产品数量减少   // 父组件接收到子组件的数据
                this.list.forEach(item=>{
                    if(item.id==value.producetId && item.num>=1){
                        item.num--
                    }
                })
            },
            numAdd(value){ // 产品数量增加
                this.list.forEach(item=>{
                    if(item.id==value.producetId){
                        item.num++
                    }
                })
            },
            producetDel(value){// 删除对应的产品
                var i=this.list.findIndex(item=>{
                    return item.id==value.producetId
                })
                this.list.splice(i,1)
            }
        }
    })
    app.component("shophead", {
        template: `
            <div class="title">{{uname}} 的购物车</div>
        `,
        props:['uname']

    })

    app.component("shopcont", {
        template: `
            <div class="shop_cont">
                <ul>
                    <li v-for="item in list">
```

```
            <img :src="item.src" class="pic">
            <div class="txt">
                <h3>{{item.name}}</h3>
                <div class="jiage">
                    <p>￥{{item.price}}</p>
                    <div class="num">
                        <input type="button" value="-" class="btn" @click="reduce(item.id)">
                        <span class="shu">{{item.num}}</span>
                        <input type="button" value="+" class="btn" @click="add(item.id)">
                    </div>
                    <i class="shan"@click="del(item.id)">&times;</i>
                </div>
            </div>
        </li>
    </ul>
</div>
`,
props:['list'],
methods:{
    reduce(id){
        this.$emit("func_reduce",{
            producetId:id
        })
    },
    add(id){
        this.$emit("func_add",{
            producetId:id
        })
    },
    del(id){
        this.$emit("func_del",{
            producetId:id
        })
    }
```

注释：
- reduce 函数，参数为产品 id
- add 函数，参数为产品 id
- del 函数，参数为产品 id
- 三个函数分别绑定自定义事件，同时传递产品的 id

```
                    }
                })

                app.component("shopfoot",{
                    template:`
                        <div class="zongjia">
                            <h2>总价：<span>{{totalPrice}}</span>元</h2>
                        </div>
                    `,
                    props:['list'],
                    computed:{
                        totalPrice(){
                            var t=0
                            this.list.forEach(item=>{
                                t=t+(item.price*item.num)

                            })
                            return t
                        }
                    }
                })
                app.mount('#container')
            }
        </script>
    </head>

    <body>
        <div id="container">
            <div class="shop">
                <shophead :uname="uname"></shophead>
                <shopcont :list="list"
                @func_reduce="numReduce($event)"
                @func_add="numAdd($event)"
                @func_del="producetDel($event)">
                </shopcont>
                <shopfoot :list="list"></shopfoot>
            </div>
```

> 三个自定义事件依次对应父组件中的三个函数，使用 $event 来接收产品的 id

```
        </div>
    </body>
</html>
```

点击"+"或者"-"按钮可以改变产品的数量,如图 7-24 所示。点击删除按钮可以删除对应的产品,如图 7-25 所示。

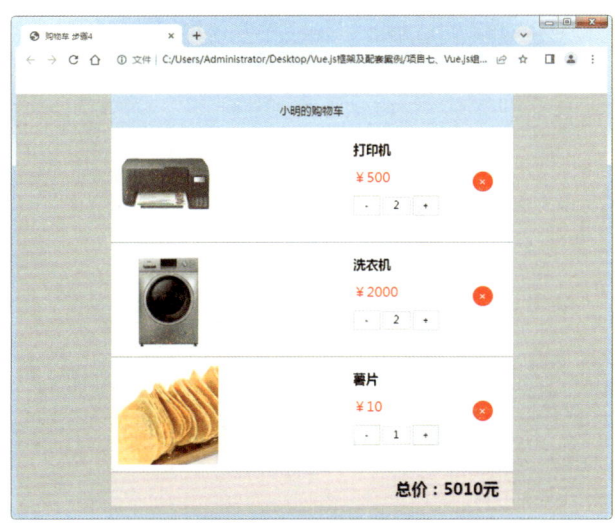

图 7-24　改变产品数量

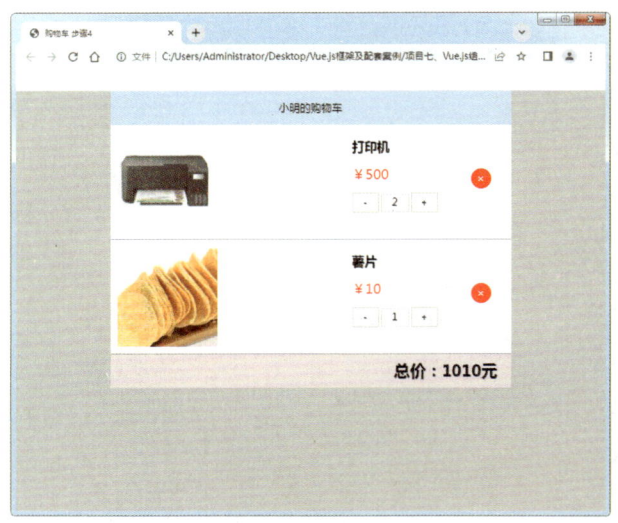

图 7-25　删除产品

课后练习题

1. 组件的内部包括多个部分,_____是模板结构,_____是该组件的方法,_____是该组件的数据。

2. 当模板部分的内容比较复杂时,推荐使用_____,模板字符串使用的符号是_____,在模板字符串中代码可以任意换行。

3. 组件名为"sayHello",在应用组件的时候要使用的标签是_____。

4. 组件的切换可以使用 Vue 提供的 component 标签,在其内部使用_____属性来指定要展示的组件的名称。

5. 父组件可以通过_____向子组件传递数据。

6. 子组件向父组件传递数值有两种方法,一种是使用_____,特点是简单直接,另一种是使用_____。

7. 自定义事件使用_____,它会绑定一个自定义事件 event,父组件通过 @event 监听并接收事件传递过来的参数。

8. 插槽的类型有_____、_____和_____等。

9. 具名插槽通过名字来进行插槽的调用。插槽使用_____来进行名称的定义,使用时通过 <template v-slot: 插槽名 ></template > 来进行调用。

第 8 章

Vue.js 路由

导 言

中国文化博大精深。中国是世界上文明发达最早的国家之一，也是世界上少有的历史文化从未间断，一直延续至今的国家，有将近4000年的有文字可考的历史。互联网时代的到来，使得中国文化在全球范围内得到了更广泛的传播和交流。中国的传统文化产品成为国家软实力的重要体现。在本章我们将学习Vue路由的相关知识，并且完成制作茶叶网站的综合案例。让我们在沁人心脾的茶香中体会中国传统文化的源远流长。

学习内容

本章一共有3节。8.1节主要讲解路由的基本应用，其中涉及的知识点包括路由的结构、Vue 2.0路由结构、嵌套路由以及链接激活的样式。8.2节主要讲解路由传参的三种方法，涉及query、params、sessionStorage的使用方法。8.3节是通过一个路由的综合案例"品茶轩"茶叶网站的制作来加深对路由的理解，提升路由的应用能力。

学习目标

1. 理解使用路由的意义。
2. 掌握Vue路由的基本语法格式。
3. 掌握嵌套路由的实现方法。
4. 掌握路由传参的多种使用方法。
5. 能够灵活应用路由来完成相应的功能。

学习重点

1. 掌握路由应用的基本语法格式。
2. 掌握路由传参的多种方法。
3. 掌握路由及组件的综合应用。

8.1 路由基础

Vue 路由的基本结构

8.1.1 Vue 路由的基本结构

在传统的 Web 开发中，点击进入新的网络地址就意味着要链接到一个新的 Web 页面，这个 Web 页面上必需的 HTML、CSS 和 JavaScript 就要重新加载一遍。在新的 Web 开发中引入了单页应用（single page web application）的概念，简称 SPA。单页应用只有一张 Web 页面，当用户与应用程序交互时，只需要动态更新这一张 Web 页面的局部内容，不需要再重新加载一次该页面的 HTML、CSS 和 JavaScript，从而提高了页面的响应速度。

由于单页应用只有一个 Web 页面，那么如何能够出现不同的网络地址呢？不同的内容对应不同的网络地址对于搜索引擎和用户来说都是必需的，例如，当用户浏览一个网页时可以直接复制或收藏当前页面的网络地址。在 Vue 中使用路由可以解决这一问题。

Vue 路由会指定不同的网络地址与 Vue 组件之间建立链接，当切换网络地址时，就会显示对应的组件。Vue 路由要使用 Vue Router，它与 Vue.js 深度集成，这使得用 Vue.js 构建单页应用变得轻而易举。

Vue 路由中使用 router-link 来定义跳转的网络地址，使用 router-view 来显示该网络地址下对应的组件。在使用路由时，组件必须使用变量的形式来进行定义。下面案例中有两个组件，一个是 Home 组件，一个是 About 组件。点击"首页"，网址变为"/home"，对应显示的是 Home 组件；点击"关于"，网址变为"/about"，对应显示的是 About 组件。具体代码如下：

```html
<!DOCTYPE html>
<html lang="en">
<head>
  <meta charset="UTF-8">
  <meta http-equiv="X-UA-Compatible" content="IE=edge">
  <meta name="viewport" content="width=device-width, initial-scale=1.0">
  <title>路由的基本结构</title>
  <style>
    a{ margin: 0 20px;}
  </style>
  <script src="./js/vue3.js"></script>
  <script src="./js/vue-router.js"></script>
  <script>
    window.onload = function () {
      const Home = { template: `<div>Home</div>` }    // 1.定义路由组件
      const About = { template: `<div>About</div>` }

      const routes = [                                // 2.定义路由规则
        { path: '/', redirect: '/home'},
        { path: '/home', component: Home },
```

```
      { path: '/about', component: About },
    ]

    const router = VueRouter.createRouter({          ← 3. 创建路由实例
      history: VueRouter.createWebHashHistory(),
      routes: routes,
    })

    const app = Vue.createApp({})

    app.use(router)                                   ← 4. 在根实例中使用路由实例
    app.mount('#container')

    }
  </script>
</head>

<body>
  <div id="container">
    <p>
      <router-link to="/home">首页</router-link>
      <router-link to="/about">关于</router-link>    ← 5. 使用 router-link 进行导航
    </p>
    <router-view></router-view>                       ← 6. 使用 router-view 显示匹配的组件
  </div>
</body>

</html>
```

从代码中可以看到路由的使用主要分为以下 6 个步骤。

（1）定义路由组件。

定义组件必须使用变量的形式来进行定义，如 const home={}，不能使用 app.component（'home',{}）的形式来进行定义。

（2）定义路由规则。

路由规则中定义 URL 地址与组件的对应关系。例如，{ path: '/home', component: Home } 的含义是当 URL 地址是 '/home' 时，在 <router-view></router-view> 的位置要显示 Home 组件。

{ path: '/', redirect: '/home'} 中 redirect 的作用是重定向。当 URL 是 '/' 时会重定向为 '/home'，类似于我们生活中的"呼叫转移"。

（3）创建路由实例。

history: VueRouter.createWebHashHistory（ ）的作用是设置 Vue 路由的模式为 Hash 模式，这种模式

是开发中的默认模式，特点是 URL 地址中会带着"#"号，如 www.000.com #/home。浏览器对这种模式的支持度很好，低版本的 IE 浏览器也支持这种模式。

routes：routes 中第二个 routes 是步骤（2）中的路由规则，因为前后两个词都相同，所以 routes: routes 可以简写为 routes。

（4）在根实例中使用路由实例。

app.use(router) 中的 router 即步骤 (3) 中创建的路由实例。

（5）使用 router-link 进行导航。

<router-link> 会默认渲染为 <a> 标签，使用通过属性 to 来指定链接的地址，例如 <router-link to="/home"> 首页 </router-link> 的含义是点击"首页"，页面的 URL 会变为 "/home"。

（6）使用 router-view 显示匹配的组件。

<router-view></router-view> 的作用相当于是一个占位符，页面的 URL 对应要显示的那个组件会在 <router-view></router-view> 的地方显示出来。

页面加载时的效果如图 8-1 所示。点击"关于"按钮，页面跳转到"/about"，同时显示 About 组件，如图 8-2 所示。<router-link> 会默认渲染为 <a> 标签，在控制台中查看的结果如图 8-3 所示，所以在 HTML 页面中直接使用 <a> 标签也是可以的，但是要注意添加上"#"号，例如：

```
<p>
    <a href="#/home"> 首页 </a>
    <a href="#/about"> 关于 </a>
</p>
```

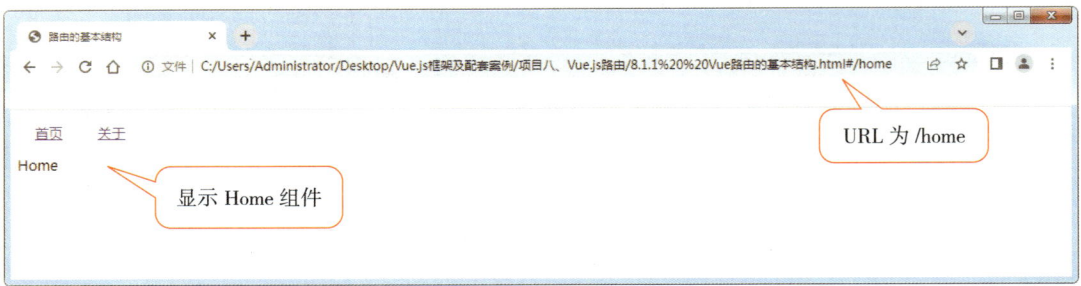

图 8-1　页面加载的初始效果

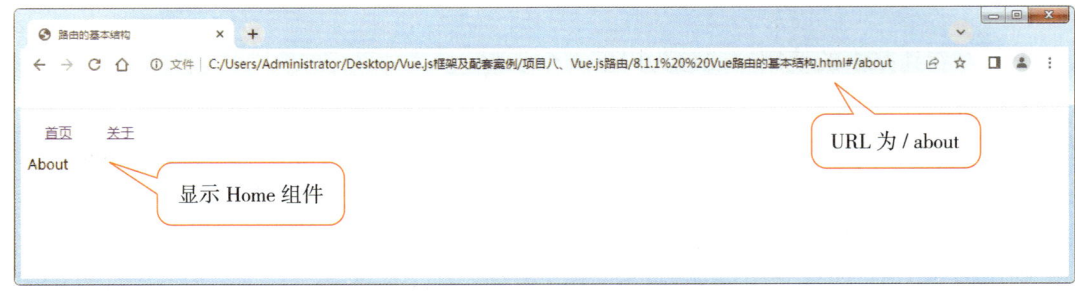

图 8-2　切换到 About 页面

图 8-3　渲染为 <a> 标签

8.1.2　使用 Vue 2.0 创建路由

上面案例中使用的是 Vue 3.0 创建的路由。由于目前企业中的一些项目使用 Vue 2.0 开发，下面介绍使用 Vue 2.0 来创建路由的方法。两个版本在创建路由时的步骤基本相同，但是细节方面还是有较大的差异，学习时要注意区分。下面案例使用 Vue 2.0 来创建路由，具体代码如下：

```
<!DOCTYPE html>
<html lang="en">
<head>
    <meta charset="UTF-8">
    <meta http-equiv="X-UA-Compatible" content="IE=edge">
    <meta name="viewport" content="width=device-width, initial-scale=1.0">
    <title>使用Vue2.0创建路由</title>
    <style>
        a{ margin: 0 20px;}
    </style>
    <script src="./js/vue.js"></script>
    <script src="./js/vue-router.js"></script>
    <script>
        window.onload = function () {
            const Home = {
                template: `<div>Home</div>`,
            }
            const About = {
                template: `<div>About</div>`
            }
```

1. 定义路由组件

```
            const routes = [
                { path: '/', redirect: '/home' },      ← 2. 定义路由规则
                { path: '/home', component: Home },
                { path: '/about', component: About },
            ]

            const router = new VueRouter({
                mode: 'hash',                          ← 3. 创建路由实例，写法与 Vue 3.0 不同
                routes: routes
            })

            new Vue({
                el: '.cont',
                data: {},
                methods: {},
                router: router                         ← 4. 注册路由，写法与 Vue 3.0 不同
            })
        }
    </script>
</head>

<body>
    <div class="cont">
        <p>
            <router-link to="/home">首页</router-link>    ← 5. 使用 router-link 进行导航
            <router-link to="/about">关于</router-link>
        </p>
        <router-view></router-view>                      ← 6. 使用 router-view 显示匹配的组件
    </div>
</body>
</html>
```

该案例效果与 8.1.1 相同，页面加载后的效果如图 8-4 所示，点击"关于"按钮后的效果如图 8-5 所示。

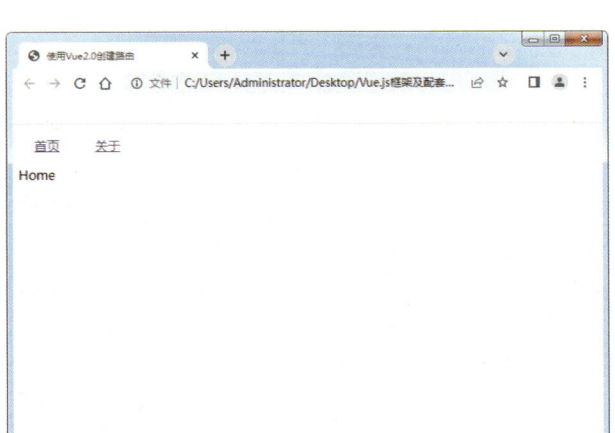

图 8-4　初始效果

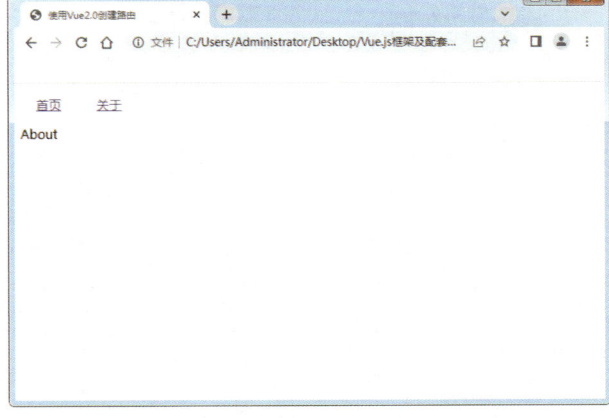

图 8-5　切换到 About 组件

8.1.3　嵌套路由

嵌套路由就是在路由里又嵌套了一个子路由。如图 8-6 所示，在用户管理页面中又有两个链接，一个是"VIP 用户"，另一个是"普通用户"。这两个链接也设置了路由，可以跳转页面。要实现这种效果就要使用嵌套路由。

嵌套子路由的关键属性是 children。children 也是一组路由，当使用 children 属性实现子路由时，子路由的 path 属性前不要带"/"，否则会以根路径开始请求。下面案例中对用户管理页面使用了嵌套路由，其代码如下：

```
<!DOCTYPE html>
<html lang="en">
<head>
  <meta charset="UTF-8">
  <meta http-equiv="X-UA-Compatible" content="IE=edge">
  <meta name="viewport" content="width=device-width, initial-scale=1.0">
  <title>Document</title>
  <style>
    a {
      margin: 0 30px;
    }
  </style>
  <script src="./js/vue3.js"></script>
  <script src="./js/vue-router.js"></script>
  <script>
    window.onload = function () {
      const Home = {
        template: `<div>Home</div>`
      }
      const About = {
```

```
    template: `<div>About</div>`
}

const User = {
  template: `
    <div>
      <h3>用户管理</h3>
      <router-link to="/user/vipuser">VIP 用户 </router-link>
      <router-link to="/user/orduser"> 普通用户 </router-link>
      <router-view></router-view>
    </div>
  `,
}

const Vip = {
  template: `<div>Vip 用户的详细内容 </div>`
}
const Ordinary = {
  template: `<div> 普通用户的详细内容 </div>`
}
const routes = [
  { path: '/', redirect: '/home' },
  { path: '/home', component: Home },
  { path: '/about', component: About },
  {
    path: '/user', component: User,
    children: [
      { path: '', component: Vip },
      { path: 'vipuser', component: Vip },
      { path: 'orduser', component: Ordinary },
    ]
  },
]

const router = VueRouter.createRouter({
  history: VueRouter.createWebHashHistory(),
  routes,
})
```

注释：
- 在 User 组件中设置 router-link 和 router-view
- Vip 组件
- Ordinary 组件
- 也可以写成 { path: '', redirect:'/user/vipuser' }
- 在 /user 内部配置子路由。这里的 path 属性不带 "/"

```
            const app = Vue.createApp({})
            app.use(router)
            app.mount('#container')
        }
    </script>
</head>
<body>
    <div id="container">
        <p>
            <router-link to="/home">首页</router-link>
            <router-link to="/about">关于我们</router-link>
            <router-link to="/user">用户管理</router-link>
        </p>
        <router-view></router-view>
    </div>
</body>
</html>
```

设置路由嵌套要在父组件中设置 <router-link></router-link> 和 <router-view></router-view>，同时要在路由规则的设置中添加 children 属性，配置子路由的对应规则。点击"用户管理"后跳转到用户管理页面，显示的是 Vip 组件，如图 8-6 所示。点击"普通用户"的效果如图 8-7 所示。

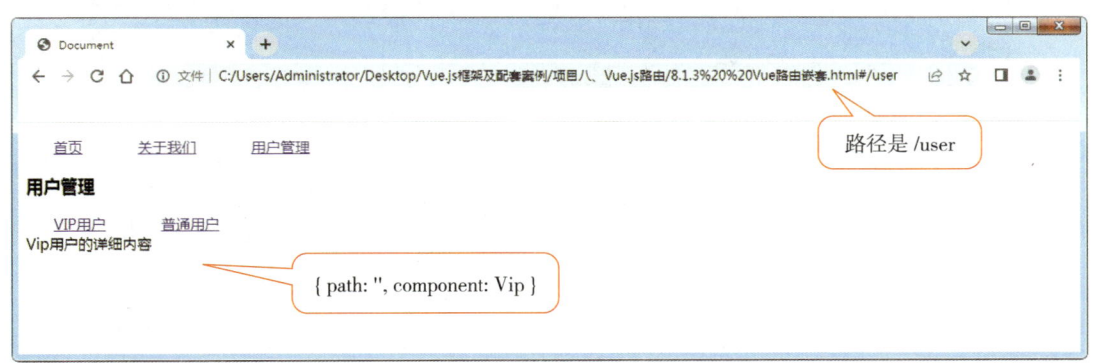

图 8-6 用户管理页面

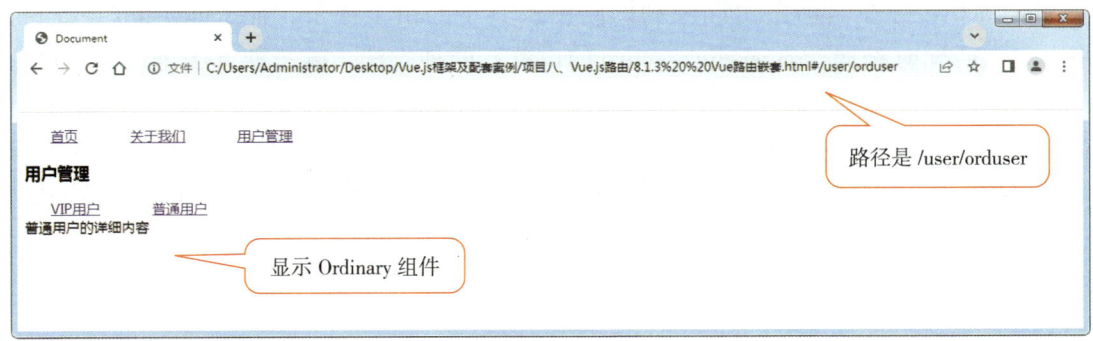

图 8-7 普通用户页面

8.1.4 设置链接激活样式

在 8.1.1 的案例中，当点击"首页"按钮时，在控制台中可以看到此时"首页"的链接中添加了两个类，如图 8-8 所示。当点击"关于"按钮时，此时"关于"的链接中添加了两个类，如图 8-9 所示。这两个类分别是 router-link-active 和 router-link-exact-active。如果页面当前的地址与链接的地址匹配上时，这个链接就处于激活状态，就会添加上这两个类。

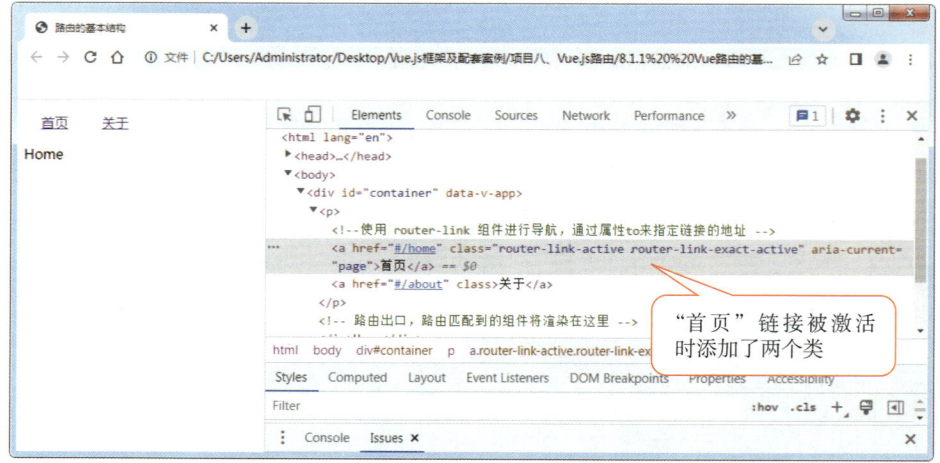

图 8-8　点击"首页"按钮添加两个类

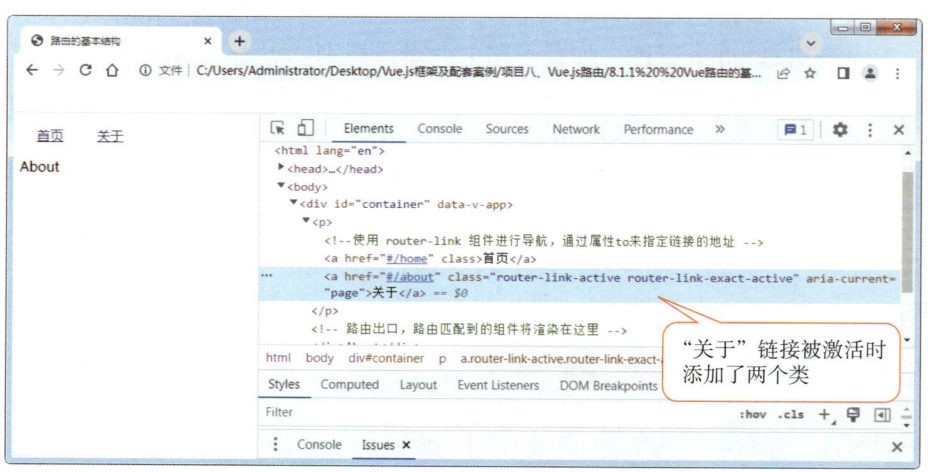

图 8-9　点击"关于"按钮添加两个类

router-link-active 会在当前路由匹配到的路由及其子路由上添加，而 router-link-exact-active 仅在当前路由完全匹配时添加。router-link-active 的设置有两种方法，第一种方法是在路由对象中使用 linkActiveClass 属性来进行全局设置，第二种方法是在 router-link 中使用 active-class 类来进行单个设置。router-link-exact-active 的设置也有两种方法，第一种方法是在路由对象中使用 linkExactActiveClass 来进行全局设置，第二种方法是在 router-link 中使用 exact-active-class 类来进行单个设置。下面案例分别对 router-link-active 和 router-link-exact-active 都进行了设置。其代码如下：

```
<!DOCTYPE html>
<html lang="en">
<head>
  <meta charset="UTF-8">
```

```html
<meta http-equiv="X-UA-Compatible" content="IE=edge">
<meta name="viewport" content="width=device-width, initial-scale=1.0">
<title>Document</title>
<style>
  a {
    width: 100px; line-height:30px; display: inline-block; margin: 0 20px; padding:
    10px; text-decoration: none; text-align: center;
  }
  .active {
    background-color: yellowgreen;
  }
  .exactActive {
    background-color: rgb(205, 50, 140);
  }

</style>
<script src="./js/vue3.js"></script>
<script src="./js/vue-router.js"></script>
<script>
  window.onload = function () {
    const Home = {
      template: `<div>Home</div>`
    }
    const About = {
      template: `<div>About</div>`
    }
    const User = {
      template: `
        <div>
          <h3>用户管理</h3>
          <router-link to="/user/vipuser">VIP用户</router-link>
          <router-link to="/user/orduser">普通用户</router-link>
          <router-view></router-view>
        </div>
        `
    }
    const Vip = {
      template: `<div>Vip用户的详细内容</div>`
    }
```

> 设置两个链接被激活后的样式

```
        const Ordinary = {
          template: `<div>普通用户的详细内容</div>`
        }
        const routes = [
          { path: '/', redirect: '/home' },
          { path: '/home', component: Home },
          { path: '/about', component: About },
          { path: '/user', component:User,
            children:[
              {path:'',component:Vip},
              {path:'vipuser',component:Vip},
              {path:'orduser',component:Ordinary},
            ]
          },
        ]

        const router = VueRouter.createRouter({
          history: VueRouter.createWebHashHistory(),
          routes,
        })

        const app = Vue.createApp({})
        app.use(router)
        app.mount('#container')
      }
    </script>
  </head>
  <body>
    <div id="container">
      <p>
        <router-link to="/home"   active-class="active" exact-active-class="exactActive" >
        首页
        </router-link>
        <router-link to="/about"   active-class="active" exact-active-class="exactActive">
        关于我们
        </router-link>
        <router-link to="/user"   active-class="active" exact-active-class="exactActive">
        用户管理
        </router-link>
```

> 路由被匹配时使用 active 类，被精准匹配时使用 exactActive 类

```
        </p>
        <router-view></router-view>
     </div>
 </body>

 </html>
```

当点击"首页""关于我们""用户管理"时，这些链接与路由完全一致，此时是完全匹配，链接会应用类 .exactActive，即背景颜色变成红色，其效果如图 8-10 所示。当点击"VIP 用户"或"普通用户"时，此时的路由为"/user/vipuser"或"/user/orduser"，与"用户管理"对应的路由"/user"不是完全匹配的。"/user"只是局部匹配了"/user/vipuser"或"/user/orduser"，此时"用户管理"会应用类 .active，即背景颜色变成黄绿色，其效果如图 8-11 所示。

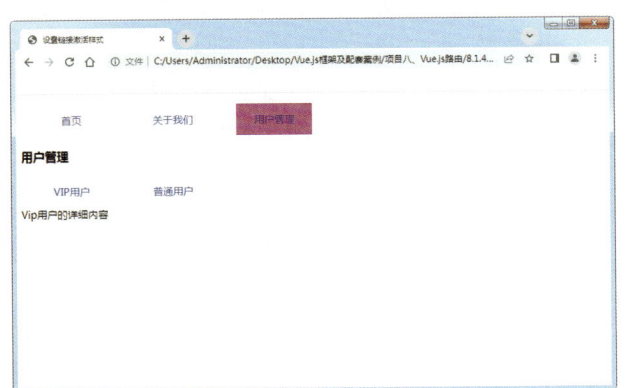
图 8-10　完全匹配

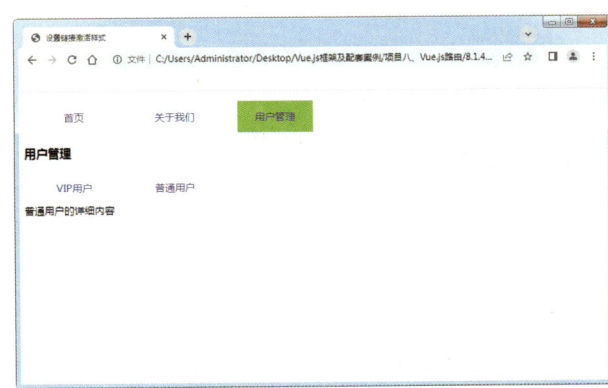
图 8-11　部分匹配

如果使用第一种方法进行全局定义，则需要在路由对象中进行声明，其代码如下：

```
const router = VueRouter.createRouter({
    history: VueRouter.createWebHashHistory(),
    routes,
    linkActiveClass: 'active',
    linkExactActiveClass:"exactActive"
})
```

因为已经在全局中进行声明，所以 router-link 中不需要再添加类，代码如下：

```
<p>
    <router-link to="/home"> 首页 </router-link>
    <router-link to="/about"> 关于我们 </router-link>
    <router-link to="/user"> 用户管理 </router-link>
</p>
```

8.2 路由传参

8.2.1 编程式导航

路由传参是指在网页 URL 跳转的过程中传递参数。在 Web 开发中，路由传参可以动态地生成链接和页面内容。通过传递参数，可以在不同页面之间进行交互，并根据参数的不同显示不同的内容，增加了网站的灵活性和个性化。

路由的链接有两种方法，一种是声明式，即上面讲到的 `<router-link to="...">` 方法；另一种是编程式方法，即使用 $router.push(…) 方法。下面案例中点击"About 页面"可以跳转到 About 页面，其代码如下：

```html
<!DOCTYPE html>
<html lang="en">
<head>
  <meta charset="UTF-8">
  <meta http-equiv="X-UA-Compatible" content="IE=edge">
  <meta name="viewport" content="width=device-width, initial-scale=1.0">
  <title>编程式导航</title>
  <script src="./js/vue3.js"></script>
  <script src="./js/vue-router.js"></script>
  <script>
    window.onload = function () {
      const Home = { template: '<div>Home</div>' }
      const About = { template: '<div>About</div>' }

      const routes = [
        { path: '/', redirect:'/home' },
        { path: '/home', component: Home },
        { path: '/about', component: About },
      ]

      const router = VueRouter.createRouter({
        history: VueRouter.createWebHashHistory(),
        routes,
      })

      const app = Vue.createApp({
        methods: {
          jumpTo() {
```

```
                this.$router.push({
                    path: '/about'
                })
            }
        }
    })

    app.use(router)
    app.mount('#container')
    }
    </script>
</head>

<body>
    <div id="container">
        <input type="button" value="About 页面" @click="jumpTo()">
        <router-view></router-view>
    </div>
</body>
</html>
```

设置跳转路径为 /about，此处也可以简写为 this.$router.push('/about')

页面加载后的效果如图 8-12 所示。点击"About 页面"后的效果如图 8-13 所示。

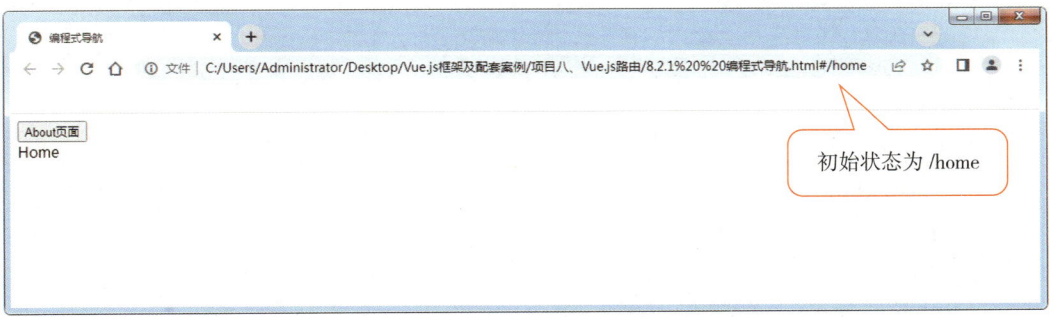

图 8-12 显示初始的 Home 页面

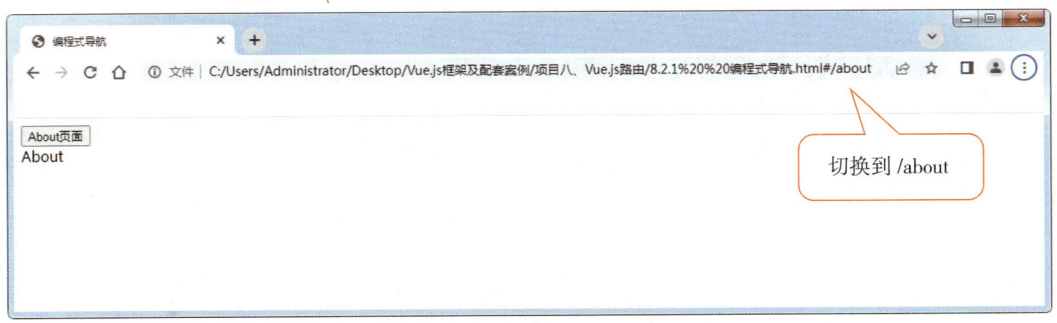

图 8-13 跳转到 About 页面

8.2.2 使用 query 路由传参

使用 $router.push( ) 在完成路由跳转的同时也可以传递参数。传递参数有两种方法，一种是使用 query，这种方法类似于表单中的 get 方法；另一种是使用 params，这种方法类似于表单中的 post 方法。下面案例在 Vue 根对象 data 中有一个变量，变量名是 id，点"About 页面"按钮，进行路由跳转，同时将 id 的值传递到 About 中。具体代码如下：

```html
<!DOCTYPE html>
<html lang="en">
<head>
  <meta charset="UTF-8">
  <meta http-equiv="X-UA-Compatible" content="IE=edge">
  <meta name="viewport" content="width=device-width, initial-scale=1.0">
  <title>使用 query 路由传参</title>
  <script src="./js/vue3.js"></script>
  <script src="./js/vue-router.js"></script>
  <script>
    window.onload = function () {
      const Home = { template: `<div>Home</div>` }
      const About = {
        template:
          `<div>
          About
          <p>获取传递过来的参数：{{this.$route.query.id}}</p>   // 接收传递过来的参数
          </div>
          `
      }

      const routes = [
        { path: '/', redirect: '/home'},
        { path: '/home', component: Home },
        { path: '/about', component: About },
      ]

      const router = VueRouter.createRouter({
        history: VueRouter.createWebHashHistory(),
        routes,
      })

      const app = Vue.createApp({
        data() {
          return {
```

```
                id: 10
            }
        },
        methods: {
            jumpTo() {
                this.$router.push({
                    path: '/about',
                    query: {
                        id: this.id
                    }
                })
            }
        }
    })
    app.use(router)
    app.mount('#container')
}
</script>
</head>
<body>
<div id="container">
    <input type="text" v-model="id">
    <input type="button" value="About 页面" @click="jumpTo()">
    <router-view></router-view>
</div>
</body>
</html>
```

（变量 id 的值为 10）

（传递参数）

页面加载后的效果如图 8-14 所示。当点击"About 页面"时，进行路由跳转，此时的 URL 携带有参数，格式为"/about?id=10"，同时在 About 组件中使用 {{this.$route.query.id}} 接收到传递过来的参数。其效果如图 8-15 所示。

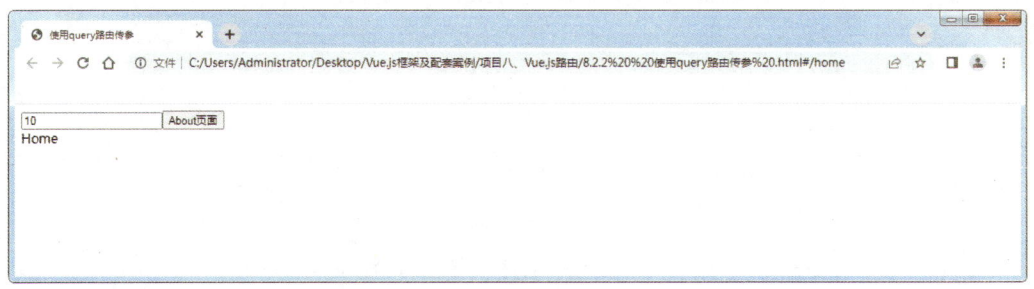

图 8-14　页面加载后的效果

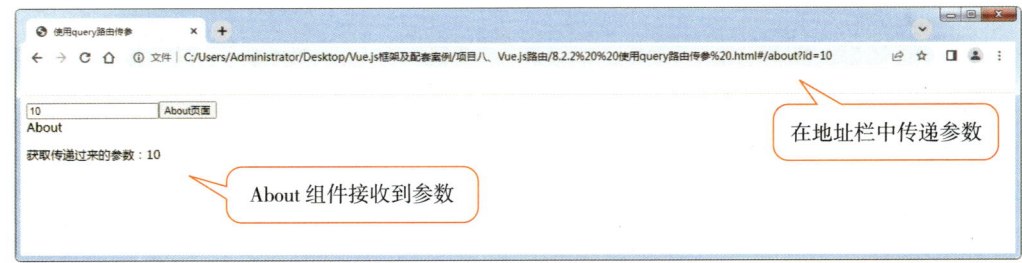

图 8-15 传递参数

8.2.3 使用 params 路由传参

使用 params 来进行路由传参时，路由对象在定义的时候必须要有 name 属性，同时使用 $router.push ( ) 时不能使用 path 属性，而要使用 name 属性，否则会进行报错。另外在定义路由规则时，path 中要使用冒号的形式来声明要传递的参数。下面案例使用 params 将 id 的值传递给 About 组件，其代码如下：

```
<!DOCTYPE html>
<html lang="en">
<head>
    <meta charset="UTF-8">
    <meta http-equiv="X-UA-Compatible" content="IE=edge">
    <meta name="viewport" content="width=device-width, initial-scale=1.0">
    <title>使用 params 路由传参</title>
    <script src="./js/vue3.js"></script>
    <script src="./js/vue-router.js"></script>
    <script>
        window.onload = function () {
            const Home = { template: `<div>Home</div>` }
            const About = {
                template:
                    `
                    <div>
                    About
                    <p>获取传递过来的参数：{{$route.params.id}}</p>
                    </div>
                    `
            }
            const routes = [
                { path: '/', redirect: '/home'},
                { path: '/home', component: Home, name: 'home' },
                { path: '/about/:id', component: About, name: 'about' },
            ]
```

添加 name 属性

:id 表示要传递 id 参数

```
            const router = VueRouter.createRouter({
                history: VueRouter.createWebHashHistory(),
                routes,
            })

            const app = Vue.createApp({
                data() {
                    return {
                        id: 10
                    }
                },
                methods: {
                    jumpTo() {
                        this.$router.push({
                            name: 'about',    // 不能使用 path 属性来定义
                            params: {
                                id: this.id
                            }
                        })
                    }
                }
            })
            app.use(router)
            app.mount('#container')
        }
    </script>
</head>
<body>
    <div id="container">
        <input type="text" v-model="id">
        <input type="button" value=" About 页面" @click="jumpTo()">
        <router-view></router-view>
    </div>
</body>
</html>
```

页面加载后的效果如图 8-16 所示。点击"About 页面"后，路由跳转到"/about"，其效果如图 8-17 所示。如果在设置 $router.push( ) 时使用 path 属性而不是 name 属性，则会报错，其效果如图 8-18 所示。

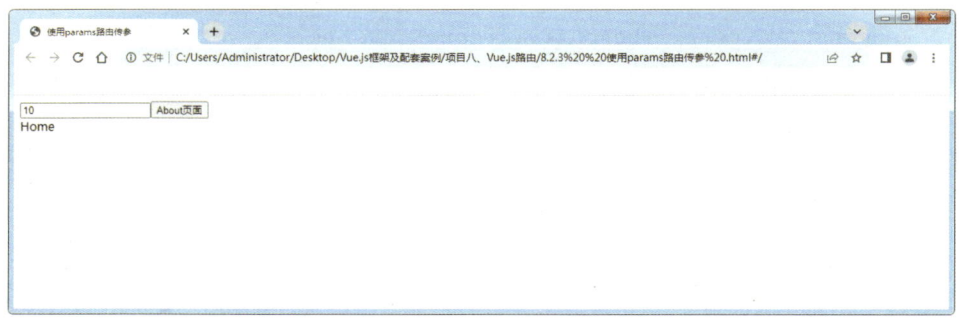

图 8-16　params 传参初始效果

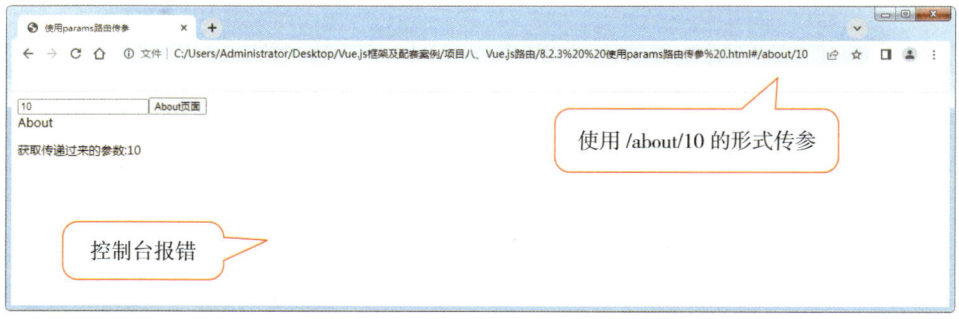

图 8-17　使用 params 传递参数

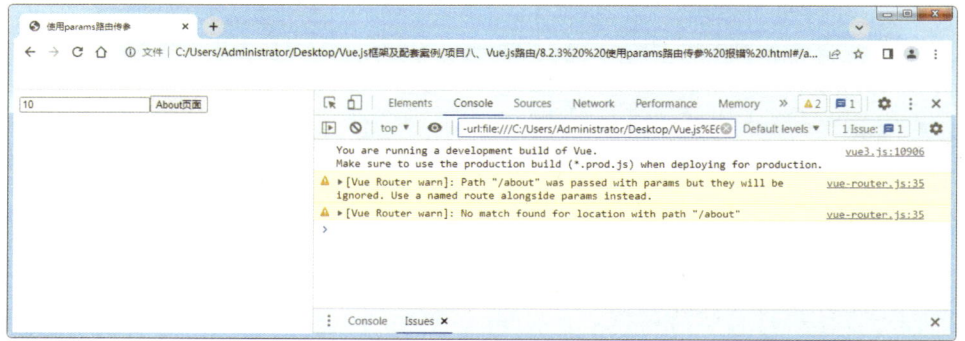

图 8-18　路由报错

8.2.4　使用 sessionStorage 传参

在 8.2.3 中使用 params 传递参数，但这样会在路径上显示出传递的参数值，对于一些敏感数据这种方法是不合适的。若使用 params 时，在定义路由规则时不使用 path: '/about/:id' 的方法而直接使用 path: '/about'，虽然参数被隐藏起来，但是当刷新页面的时候参数会丢失，其效果如图 8-19 所示。

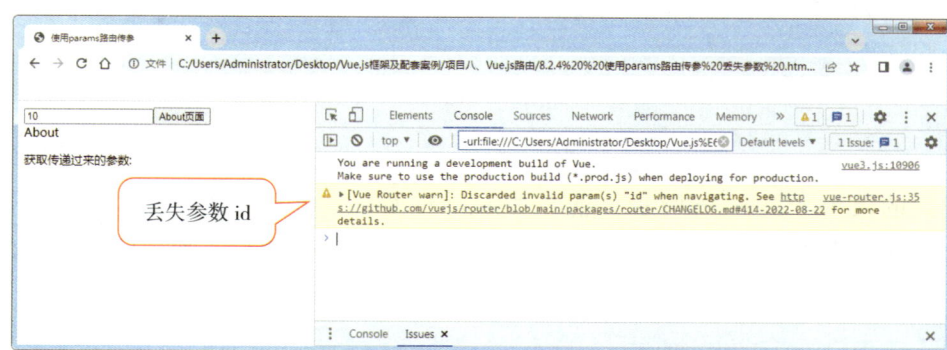

图 8-19　丢失参数

为了解决这一问题可以使用 Vuex 或者 sessionStorage。这里介绍 sessionStorage 的使用方法。

sessionStorage 是一个会话存储对象，用于临时保存同一窗口的数据，在关闭窗口之后将会删除这些数据。在 JavaScript 中可通过 window.sessionStorage 调用此对象。seesionStorage 的存储方式采用 key、value 的方式。value 的值必须为字符串类型。下面案例通过 seesionStorage 的方法来进行传参数。首先在点击"About 页面"时使用 sessionStorage 的 setItem 方法来存储一组键值对，然后再使用 sessionStorage 的 getItem 方法来获取对应的值。案例代码如下：

```html
<!DOCTYPE html>
<html lang="en">
<head>
    <meta charset="UTF-8">
    <meta http-equiv="X-UA-Compatible" content="IE=edge">
    <meta name="viewport" content="width=device-width, initial-scale=1.0">
    <title>使用 params 路由传参</title>
    <script src="./js/vue3.js"></script>
    <script src="./js/vue-router.js"></script>
    <script>
        window.onload = function () {
            const Home = { template: `<div>Home</div>` }
            const About = {
                template:
                `
                <div>
                About
                <p>获取传递过来的参数:{{id}}</p>

                </div>
                `,
                data(){
                    return {
                        id:''
                    }
                },
                methods:{
                    getId(){
                        this.id=window.sessionStorage.getItem('id')  // 获取在 setItem 中存储的那个 id 值
                    }
                },
                mounted(){  // 组件实例被挂载到 DOM 后就执行 getId() 函数
                    this.getId()
```

```
            }
        }

        const routes = [
            { path: '/', redirect: '/home'},
            { path: '/home', component: Home, name: 'home' },
            { path: '/about', component: About, name: 'about' },
        ]

        const router = VueRouter.createRouter({
            history: VueRouter.createWebHashHistory(),
            routes,
        })

        const app = Vue.createApp({
            data() {
                return {
                    id: 10
                }
            },
            methods: {
                jumpTo() {
                    this.$router.push({
                        name:'about',
                    })

                    window.sessionStorage.setItem("id", JSON.stringify(this.id));
                }
            },
        })

        app.use(router)
        app.mount('#container')
    }
    </script>
</head>
<body>
    <div id="container">
        <input type="text" v-model="id">
        <input type="button" value="About 页面" @click="jumpTo()">
```

> 存储一组键值对，JSON.stringify 的作用是将 JSON 对象变成字符串形式

```
        <router-view></router-view>
    </div>
</body>
</html>
```

在控制台 Application 下的 Storage 的 Session Storage 中可以看到，页面初始化的时候，Session Storage 中是没有值的，如图 8-20 所示。当点击"About 页面"后，Session Storage 中出现一组键值对，如图 8-21 所示。

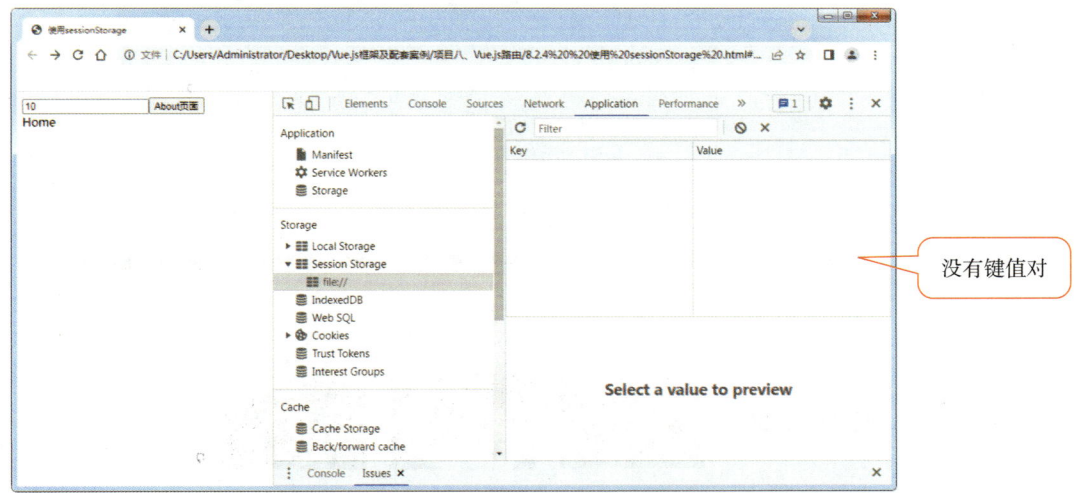

图 8-20　页面初始化时没有键值对

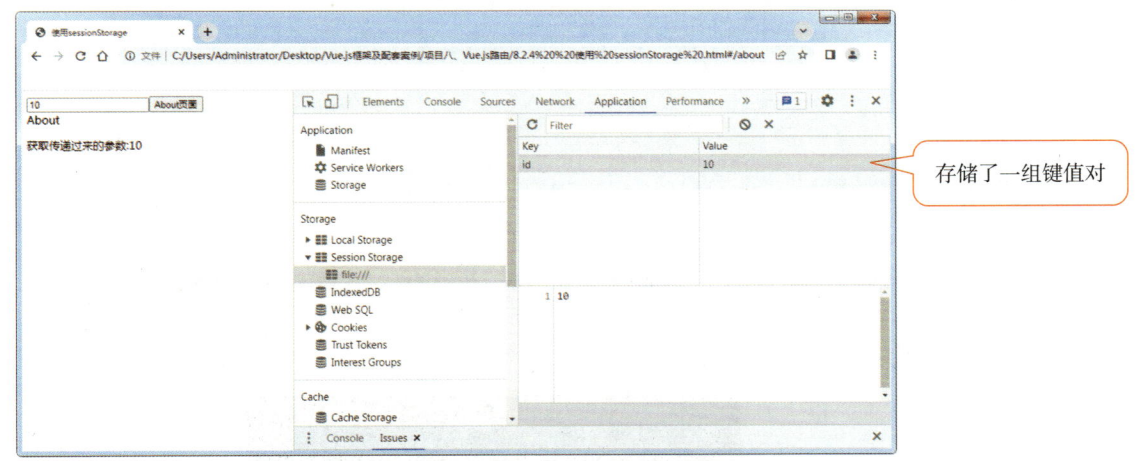

图 8-21　Session Storage 存储键值对

8.3　综合实训：茶叶网站的制作

8.3.1　项目描述

制作茶叶网站可以使用 Bootstrap 前端框架来完成样式的设置，使用路由来完成页面的跳转。网站导航分为三个部分：首页、茶叶分类和关于，其效果如图 8-22 所示。茶叶分类中有三种茶叶类型，具体效果如图 8-23 所示。点击具体的茶叶类型又可以跳转到详情页面，其效果如图 8-24 所示。

图 8-22　茶叶网站的首页

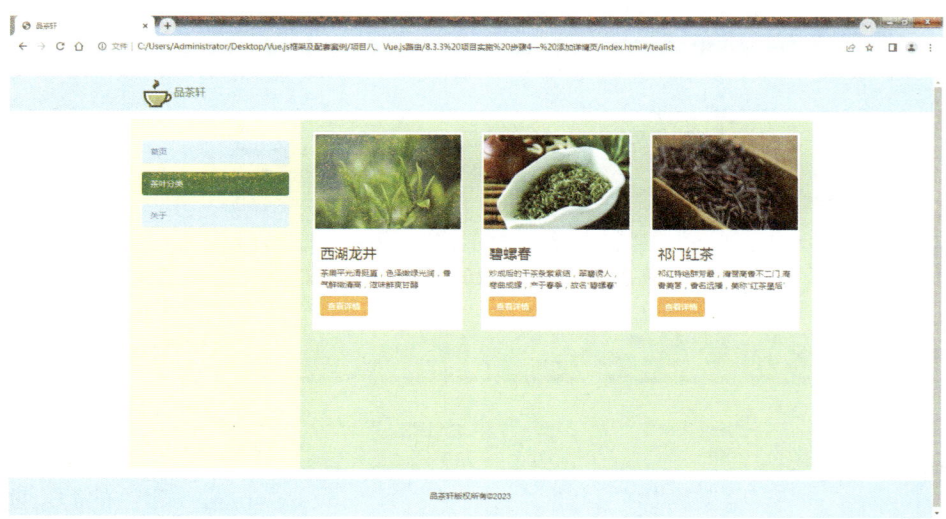

图 8-23　茶叶分类

图 8-24　茶叶详情

8.3.2 项目分析

"首页""茶叶分类""关于"是三个组件,使用路由设置,让这三个路由组件根据路由规则在右侧进行显示。如图 8-23 所示,点击"查看详情"可以进行路由的跳转,进入详情页,所以详情页也是一个组件,使用编程式导航,同时要将茶叶的 id 值传递过去,详情页组件接收 id 值,将对应的茶叶详情显示出来。页面的数据放置在 Vue 根对象中,子组件与父组件要使用组件的相关知识进行数据传递。

8.3.3 项目实施

项目实施步骤 1:完成静态界面

静态界面使用 Bootstrap 框架进行设置,使用 Bootstrap 框架可以通过调用各种样式来快速搭建响应式布局的网站。对 Bootstrap 框架不熟悉的读者可以到官网 https://www.bootcss.com/ 中查阅学习,此处不再赘述。该项目使用的是 Bootstrap V3 版本,对应的学习网址为 https://v3.bootcss.com/。除了使用 Bootstrap 框架的样式,有个别的样式也需要单独设置,具体代码如下:

```html
<!DOCTYPE html>
<html lang="en">
<head>
    <meta charset="UTF-8">
    <meta http-equiv="X-UA-Compatible" content="IE=edge">
    <meta name="viewport" content="width=device-width, initial-scale=1.0">
    <title>品茶轩</title>
    <link rel="stylesheet" href="./css/bootstrap.min.css">
    <link rel="stylesheet" href="./css/tea.css">
</head>
<body>
    <div id="app">
        <div class="header bg-info">
            <div class="container">
                <img src="./images/logo.png" class="logo">
                <span class="h4 text-success">品茶轩</span>
            </div>
        </div>


        <div class="container">
            <div class="row">
                <div class="aside col-md-3  bg-warning">   <!-- 页面左侧 -->
                    <ul class="nav nav-pills nav-stacked">
                        <li><a href="#" class="bg-info">首页</a></li>
                        <li><a href="#" class="bg-info">茶叶分类</a></li>
```

```html
                    <li><a href="#" class="bg-info">关于</a></li>
                </ul>
            </div>
            <div class="main col-md-9 bg-success">   <!-- 页面右侧 -->
                <div class="home">
                    <img src="./images/banner.jpg" alt="" class="img-responsive">
                    <p>茶叶，俗称茶，一般包括茶树的叶子和芽。茶叶成分有儿茶素、茶多酚、茶氨酸等。适量喝茶有益健康。茶叶制成的茶饮料，是世界三大饮料之一。
                    </p>
                    <p>茶叶源于中国，茶叶最早是被作为祭品使用的。但从春秋后期就被人们作为菜食，在西汉中期发展为药用，西汉后期才发展为宫廷高级饮料，普及民间作为普通饮料那是西晋以后的事。发现最早人工种植茶叶的遗迹在浙江余姚的田螺山遗址，已有6000多年的历史。
                    </p>
                </div>
            </div>
        </div>

        <div class="footer bg-info clearfix">
            <p class="text-center">品茶轩版权所有&copy;2023</p>
        </div>
    </div>
</body>
</html>
```

tea.css 对应的代码如下：

```css
.header{ margin-bottom: 15px;}
.aside{ padding: 20px ; height: 600px;}
.aside li{ margin: 15px 0 }
.main{ height: 600px;padding: 20px;}
.footer{ line-height: 60px; margin-top: 15px;}
```

页面的效果如图 8-25 所示。

图8-25　茶叶网站的首页

项目实施步骤2：设置路由

定义三个组件"Home""Tealist""About"，设置路由规则，同时将页面右侧的位置替换成<router-view></router-view>，其代码如下：

```
<!DOCTYPE html>
<html lang="en">
<head>
    <meta charset="UTF-8">
    <meta http-equiv="X-UA-Compatible" content="IE=edge">
    <meta name="viewport" content="width=device-width, initial-scale=1.0">
    <title>品茶轩</title>
    <link rel="stylesheet" href="./css/bootstrap.min.css">
    <link rel="stylesheet" href="./css/tea.css">
    <script src="./js/vue3.js"></script>
    <script src="./js/vue-router.js"></script>
    <script>
        window.onload = function () {
            const app = Vue.createApp({})

            const Home = {           //Home 组件，对应"首页"
                template:
                    `<div class="home">
                        <img src="./images/banner.jpg" alt="" class="img-responsive">
                        <p>茶叶，俗称茶，一般包括茶树的叶子和芽。茶叶成分有儿茶素、茶多酚、茶氨酸等。适量喝茶有益健康。茶叶制成的茶饮料，是世界三大饮料之一。
                        </p>
```

```
            <p>茶叶源于中国，茶叶最早是被作为祭品使用的。但从春秋后期就被人们
            作为菜食，在西汉中期发展为药用，西汉后期才发展为宫廷高级饮料，普及
            民间作为普通饮料那是西晋以后的事。发现最早人工种植茶叶的遗迹在浙江
            余姚的田螺山遗址，已有6000多年的历史。
            </p>
        </div>
    `
}
const Tealist = {                 ← Tealist 组件，对应"茶叶分类"
    template:
    ` <div class="tealist row">
            <div class="col-md-4">
            <div class="thumbnail">
                <img src="./images/1.jpg" >
                <div class="caption">
                <h3> 西湖龙井 </h3>
                <p> 茶扁平光滑挺直，色泽嫩绿光润，香气鲜嫩清高，滋味鲜爽甘醇 </p>
                <p><button href="#" class="btn btn-warning">
                    查看详情 </button> </p>
                </div>
            </div>
            </div>

            <div class="col-md-4">
            <div class="thumbnail">
                <img src="./images/2.jpg" >
                <div class="caption">
                <h3> 碧螺春 </h3>
                <p> 炒成后的干茶条索紧结，翠碧诱人，卷曲成螺，产于春季，故名"
                碧螺春"</p>
                <p><button href="#" class="btn btn-warning">
                    查看详情 </button> </p>
                </div>
            </div>
            </div>

            <div class="col-md-4">
            <div class="thumbnail">
                <img src="./images/3.jpg" >
                <div class="caption">
```

```
                    <h3>祁门红茶</h3>
                    <p>祁红特绝群芳最，清誉高香不二门</p>
                    <p><button href="#" class="btn btn-warning">
                      查看详情</button></p>
                  </div>
               </div>
             </div>
          </div>
          `
    }

    const About = {          // About 组件，对应"关于"
        template:
          `<div class="about">
              <p>普及茶文化知识是我们的责任和使命，因为茶文化是我们民族独有的文
              化瑰宝，也是我们国家的重要软实力。通过普及茶文化知识，可以增强人们
              对传统文化的认同感和自豪感，促进文化传承和创新。茶文化也是中外交流
              的重要桥梁，通过茶文化的交流可以增进不同国家和地区的友谊和了解。
              </p>
              <h3>联系我们</h3>
              <form>
                 <div class="form-group">
                   <label for="name">姓名</label>
                   <input type="text" class="form-control" id="name"
                    placeholder="请输入姓名">
                 </div>
                 <div class="form-group">
                     <label for="idea">建议</label>
                     <input type="text" class="form-control" id="idea"
                     placeholder="请输入建议">
                 </div>
                 <button type="submit" class="btn btn-default">提交</button>
              </form>
           </div>
           `
    }

    const routes = [         // 设置路由规则
```

```
                { path: '/', redirect:'/home' },
                { path: '/home', component: Home },
                { path: '/tealist', component: Tealist },
                { path: '/about', component: About },
            ]

            const router = VueRouter.createRouter({      ← 设置路由对象
                history: VueRouter.createWebHashHistory(),
                routes,
            })
            app.use(router)
            app.mount('#app')
        }
    </script>

</head>

<body>
    <div id="app">
        <div class="header bg-info">
            <div class="container">
                <img src="./images/logo.png" class="logo">
                <span class="h4 text-success">品茶轩</span>
            </div>
        </div>

        <div class="container">
            <div class="row">
                <div class="aside col-md-3  bg-warning">
                    <ul class="nav nav-pills nav-stacked">
                        <li><a href="#" class="bg-info">首页</a></li>
                        <li><a href="#" class="bg-info">茶叶分类</a></li>
                        <li><a href="#" class="bg-info">关于</a></li>
                    </ul>
                </div>
                <div class="main col-md-9  bg-success">
                    <router-view></router-view>      ← 页面右侧使用 router-view
                </div>                                  动态显示路由对应的组件
            </div>
```

```
            </div>
            
            <div class="footer bg-info clearfix">
                <p class="text-center"> 品茶轩版权所有 &copy;2023</p>
            </div>
        </div>
    </body>
</html>
```

项目实施步骤3：添加数据

在步骤2中组件中的所有内容都是固定的，在步骤3中将导航信息的数据以及茶叶分类的数据定义在Vue根对象中，左侧导航以及Tealist组件从Vue根对象中获取数据，具体代码如下：

```
<!DOCTYPE html>
<html lang="en">

<head>
    <meta charset="UTF-8">
    <meta http-equiv="X-UA-Compatible" content="IE=edge">
    <meta name="viewport" content="width=device-width, initial-scale=1.0">
    <title> 品茶轩 </title>
    <link rel="stylesheet" href="./css/bootstrap.min.css">
    <link rel="stylesheet" href="./css/tea.css">
    <script src="./js/vue3.js"></script>
    <script src="./js/vue-router.js"></script>
    <script>
        window.onload = function () {
            const app = Vue.createApp({
                data(){
                    return {
                        nav:[          定义导航信息的数据
                            {
                                label:'首页',
                                path:'/home'
                            },
                            {
                                label:'茶叶分类',
                                path:'/tealist'
                            },
                            {
                                label:'关于',
```

```
            path:'/about'
        },
    ],
    tealist:[          定义茶叶分类的数据
        {
            id:0,
            name:'西湖龙井',
            type:'绿茶',
            place:'浙江省杭州市西湖',
            feature:'茶扁平光滑挺直，色泽嫩绿光润，香气鲜嫩清高，滋味鲜爽甘醇',
            detail:'特级西湖龙井，叶底细嫩呈朵。"院外风荷西子笑，明前龙井女儿红。"西湖龙井茶与西湖一样，是人、自然、文化三者的完美结晶，是西湖地域文化的重要载体。',
            smallImg:'./images/1.jpg',
            bigImg:'./images/big1.jpg'
        },
        {
            id:1,
            name:'碧螺春',
            type:'绿茶',
            place:'江苏省苏州市太湖洞庭山',
            feature:'炒成后的干茶条索紧结，翠碧诱人，卷曲成螺，产于春季，故名"碧螺春"',
            detail:'唐朝时就被列为贡品，古人们又称碧螺春为"工夫茶"。 此茶冲泡后杯中白云翻滚，清香袭人，是中国的名茶。主要工序为杀青、揉捻、搓团显毫、炒青。',
            smallImg:'./images/2.jpg',
            bigImg:'./images/big2.jpg'
        },
        {
            id:2,
            name:'祁门红茶',
            type:'红茶',
            place:'安徽省祁门一带',
            feature:'祁红特绝群芳最，清誉高香不二门，高香美誉，香名远播，美称"红茶皇后"',
            detail:'茶叶原料选用当地的中叶、中生种茶树"槠叶种"（又名祁门种）制作，是中国历史名茶，著名红茶精品。祁门红茶是红茶中的极品，高香美誉，香名远播，享有盛誉。',
```

```
                smallImg:'./images/3.jpg',
                bigImg:'./images/big3.jpg'
            },

        ]
    }

    }
})

const Home = { template:
    `<div class="home">
            <img src="./images/banner.jpg" alt="" class="img-responsive">
            <p>茶叶,俗称茶,一般包括茶树的叶子和芽。茶叶成分有儿茶素、茶多酚、茶氨酚等。适量喝茶有益健康。茶叶制成的茶饮料,是世界三大饮料之一。</p>
            <p>茶叶源于中国,茶叶最早是被作为祭品使用的。但从春秋后期就被人们作为菜食,在西汉中期发展为药用,西汉后期才发展为宫廷高级饮料,普及民间作为普通饮料那是西晋以后的事。发现最早人工种植茶叶的遗迹在浙江余姚的田螺山遗址,已有6000多年的历史。
            </p>
    </div>
    `
}

const Tealist = {
    template:
        `<div class="tealist row">
            <div class="col-md-4" v-for="item in tealist">
            <div class="thumbnail">
                <img :src="item.smallImg" >
                <div class="caption">
                <h3>{{item.name}}</h3>
                <p>{{item.feature}}</p>
                <p><button class="btn btn-warning">查看详情</button>
                </p>
                </div>
            </div>
```

```
            </div>
        </div>
        `,
        props:['tealist']   ← 在 Tealist 组件中使用 props
}

const About = {
    template:
        `
        <div class="about">
            <p>普及茶文化知识是我们的责任和使命，因为茶文化是我们民族独有的文
            化瑰宝，也是我们国家的重要软实力。通过普及茶文化知识，可以增强人们
            对传统文化的认同感和自豪感，促进文化传承和创新。茶文化也是中外交流
            的重要桥梁，通过茶文化的交流可以增进不同国家和地区的友谊和了解。</
            p>
            <h3>联系我们</h3>
            <form>
                <div class="form-group">
                    <label for="name">姓名</label>
                    <input type="text" class="form-control" id="name"
                    placeholder="请输入姓名">
                </div>
                <div class="form-group">
                    <label for="idea">建议</label>
                    <input type="text" class="form-control" id="idea"
                    placeholder="请输入建议">
                </div>
                <button type="submit" class="btn btn-default">提交
                </button>
            </form>
        </div>
        `
}

const routes = [
    { path: '/', redirect:'/home' },
    { path: '/home', component: Home },
    { path: '/tealist', component: Tealist},
    { path: '/about', component: About },
```

```
            ]
            const router = VueRouter.createRouter({
                history: VueRouter.createWebHashHistory(),
                routes,
            })

            app.use(router)
            app.mount('#app')
        }
    </script>
</head>
<body>
    <div id="app">
        <div class="header bg-info">
            <div class="container">
                <img src="./images/logo.png" class="logo">
                <span class="h4 text-success">品茶轩</span>
            </div>
        </div>

        <div class="container">
            <div class="row">
                <div class="aside col-md-3 bg-warning">
                    <ul class="nav nav-pills nav-stacked">
                        <li v-for="(value,index) in nav">
                            <router-link :to="value.path" class="bg-info"
                            active-class="active">{{value.label}}
                            </router-link>
                        </li>
                    </ul>
                </div>
                <div class="main col-md-9 bg-success">
                    <router-view :tealist="tealist"></router-view>
                </div>
            </div>
        </div>

        <div class="footer bg-info clearfix">
```

使用 v-for 循环导航数据

链接激活后的样式

从 Vue 根对象中传递数据到 Tealist 组件

```
                <p class="text-center">品茶轩版权所有&copy;2023</p>
            </div>
        </div>
    </body>
</html>
```

导航部分添加了链接激活后的样式 active，为了覆盖 Bootstrap 框架有已有的样式，需要在样式中添加"!important"来提高 active 中样式的优先级。tea.css 中要新增 active 样式，其代码如下：

```
.header{ margin-bottom: 15px;}
.aside{ padding: 20px ; height: 600px;}
.aside li{ margin: 15px 0 }
.main{ height: 600px;padding: 20px;}
.footer{ line-height: 60px; margin-top: 15px;}
.active{ background-color: green !important; color: #FFFFFF !important;}
```
定义 active 样式

页面效果如图 8-26 所示。

图 8-26　组件获取数据

项目实施步骤 4：添加详情页

在步骤 4 中要完成点击"查看详情"按钮跳转到相应的详情页。点击"查看详情"按钮要将茶叶的 id 通过路由进行传递，在详情页面中接收传递进来的 id，将数据渲染到详情页组件中，具体代码如下：

```
<!DOCTYPE html>
<html lang="en">
    <head>
        <meta charset="UTF-8">
        <meta http-equiv="X-UA-Compatible" content="IE=edge">
        <meta name="viewport" content="width=device-width, initial-scale=1.0">
        <title>品茶轩</title>
        <link rel="stylesheet" href="./css/bootstrap.min.css">
        <link rel="stylesheet" href="./css/tea.css">
```

项目实施步骤 4：添加详情页

```
<script src="./js/vue3.js"></script>
<script src="./js/vue-router.js"></script>
<script>
    window.onload = function () {
        const app = Vue.createApp({
            data(){
                return {
                    nav:[
                        {
                            label:'首页',
                            path:'/home'
                        },
                        {
                            label:'茶叶分类',
                            path:'/tealist'
                        },
                        {
                            label:'关于',
                            path:'/about'
                        },
                    ],
                    tealist:[
                        {
                            id:0,
                            name:'西湖龙井',
                            type:'绿茶',
                            place:'浙江省杭州市西湖',
                            feature:'茶扁平光滑挺直，色泽嫩绿光润，香气鲜嫩清高，滋味鲜爽甘醇',
                            detail:'特级西湖龙井，叶底细嫩呈朵。"院外风荷西子笑，明前龙井女儿红。"西湖龙井茶与西湖一样，是人、自然、文化三者的完美结晶，是西湖地域文化的重要载体。',
                            smallImg:'./images/1.jpg',
                            bigImg:'./images/big1.jpg'
                        },
                        {
                            id:1,
                            name:'碧螺春',
                            type:'绿茶',
```

```
                place:'江苏省苏州市太湖洞庭山',
                feature:'炒成后的干茶条索紧结,翠碧诱人,卷曲成螺,产于春季,
                故名"碧螺春"',
                detail:'唐朝时就被列为贡品,古人们又称碧螺春为"工夫茶"。 此
                茶冲泡后杯中白云翻滚,清香袭人,是中国的名茶。主要工序为杀青、
                揉捻、搓团显毫、炒青。',
                smallImg:'./images/2.jpg',
                bigImg:'./images/big2.jpg'
            },
            {
                id:2,
                name: '祁门红茶',
                type: '红茶',
                place: '安徽省祁门一带',
                feature:'祁红特绝群芳最,清誉高香不二门,高香美誉,香名远播,
                美称"红茶皇后"',
                detail:'茶叶原料选用当地的中叶、中生种茶树"槠叶种"(又名祁
                门种)制作,是中国历史名茶,著名红茶精品。祁门红茶是红茶中的极
                品,高香美誉,香名远播,享有盛誉。',
                smallImg:'./images/3.jpg',
                bigImg:'./images/big3.jpg'
            },
        ]
    }
})

const Home = { template:
    `<div class="home">
        <img src="./images/banner.jpg" alt="" class="img-
        responsive">
        <p>茶叶,俗称茶,一般包括茶树的叶子和芽。茶叶成分有儿茶素、茶多酚、
        茶氨酸等。适量喝茶有益健康。茶叶制成的茶饮料,是世界三大饮料之一。
        </p>
        <p>茶叶源于中国,茶叶最早是被作为祭品使用的。但从春秋后期就被人们
        作为菜食,在西汉中期发展为药用,西汉后期才发展为宫廷高级饮料,普及
        民间作为普通饮料那是西晋以后的事。发现最早人工种植茶叶的遗迹在浙江
        余姚的田螺山遗址,已有6000多年的历史。
```

```
            </p>
        </div>
    `}

const Tealist = {
    template:
        `
        <div class="tealist row">
            <div class="col-md-4" v-for="item in tealist">
                <div class="thumbnail">
                    <img :src="item.smallImg" >
                    <div class="caption">
                    <h3>{{item.name}}</h3>
                    <p>{{item.feature}}</p>
                    <p><button  class="btn btn-warning"
                        @click="toDetail (item.id)">查看详情</button> </p>
                    </div>
                </div>
            </div>
        </div>
        `,
    props:['tealist'],
    methods:{
        toDetail(teaId){
            this.$router.push({
                path:'/detail',
                query:{
                    id:teaId
                }
            })
        }
    }
}

const About = {
    template:
        `
        <div class="about">
```

将茶叶的 id 值作为函数参数

使用 query 进行路由传参

```
            <p>普及茶文化知识是我们的责任和使命,因为茶文化是我们民族独有的文化
            瑰宝,也是我们国家的重要软实力。通过普及茶文化知识,可以增强人们对传
            统文化的认同感和自豪感,促进文化传承和创新。茶文化也是中外交流的重要
            桥梁,通过茶文化的交流可以增进不同国家和地区的友谊和了解。</p>
            <h3>联系我们</h3>
            <form>
                <div class="form-group">
                <label for="name">姓名</label>
                <input type="text" class="form-control" id="name"
                placeholder="请输入姓名">
                </div>
                <div class="form-group">
                    <label for="idea">建议</label>
                    <input type="text" class="form-control" id="idea"
                    placeholder="请输入建议">
                </div>
                <button type="submit" class="btn btn-default">提交
                </button>
            </form>
        </div>
        `
    }

    const Detail = {          // 定义茶叶详情组件
        template:
            `
            <div class="detail">
                <img :src="tea.bigImg" class="img-responsive">
                <h3>名称:{{tea.name}}</h3>
                <p><b>类型:</b>{{tea.type}}</p>
                <p><b>产地:</b>{{tea.place}}</p>
                <p><b>形态:</b>{{tea.feature}}</p>
                <p><b>特色:</b>{{tea.detail}}</p>
            </div>
            `,
        data(){
            return {
                tea:'',
                id:'',
```

```
                    }
                },
                methods:{
                    getDetail(){
                       var result= this.tealist.filter(item=>{
                            return item.id==this.id
                        })
                        this.tea=result[0]
                    }
                },
                mounted(){
                    this.id=this.$route.query.id
                    this.getDetail()
                },
                props:['tealist']
            }

            const routes = [
                { path: '/', redirect:'/home' },
                { path: '/home', component: Home },
                { path: '/tealist', component: Tealist},
                { path: '/about', component: About },
                { path: '/detail', component: Detail },
            ]

            const router = VueRouter.createRouter({
                history: VueRouter.createWebHashHistory(),
                routes,
            })

            app.use(router)
            app.mount('#app')
        }
    </script>
</head>
<body>
    <div id="app">
        <div class="header bg-info">
            <div class="container">
```
注释:
- 使用filter方法返回符合id值要求的数组赋值给变量tea
- 获取路由传参中的id值
- 定义在getDetail()中使用tealist
- 添加Detail组件的路由规则

```html
                    <img src="./images/logo.png" class="logo">
                    <span class="h4 text-success">品茶轩</span>
                </div>
            </div>

            <div class="container">
                <div class="row">
                    <div class="aside col-md-3  bg-warning">
                        <ul class="nav nav-pills nav-stacked">
                            <li v-for="(value,index) in nav" >
                                <router-link :to="value.path"  class="bg-info" active-class="active">{{value.label}}</router-link>
                            </li>
                        </ul>
                    </div>
                    <div class="main col-md-9  bg-success">
                        <router-view :tealist="tealist"></router-view>
                    </div>
                </div>
            </div>

            <div class="footer bg-info clearfix">
                <p class="text-center">品茶轩版权所有&copy;2023</p>
            </div>
        </div>
    </body>
</html>
```

点击"查看详情"按钮完成路由的跳转，同时显示详情页组件 Detail，其效果如图 8-27 所示。

图 8-27　详情页

课后练习题

1. 单页应用（single page web application），简称_____。

2. 路由中使用_____来定义跳转的网络地址，使用_____来显示该网络地址下对应的组件。在使用路由时，组件必须使用_____的形式来进行定义。

3. Vue 路由的模式为 Hash 模式时，URL 地址中会带有的符号为_____。

4. <router-link> 默认被渲染为_____标签。

5. 嵌套子路由的关键属性是_____。

6. 会在当前路由匹配到的路由及其子路由上时添加的类是_____，仅在当前路由完全匹配时添加的类是_____。

7. 使用 $router.push( ) 传递参数有两种方法，一种是使用_____，这种方法类似于表单中的 get 方法；另一种方法是使用_____，这种方法类似于表单中的 post 方法。

8. sessionStorage 是一个会话存储对象，用于临时保存同一窗口的数据，在 JavaScript 中可通过_____调用此对象。

第 9 章

Vue CLI

导言

实现中华民族伟大复兴，就是中华民族近代以来最伟大的梦想。这个梦想需要每一个中华儿女共同努力、共同奋斗。在这个奋进的时代，祖国给予了我们巨大的发展空间。在本章中我们要学习 Vue CLI。Vue CLI 是项目搭建的平台，而祖国就是我们追求梦想的平台。春风中，让我们播撒梦想的种子。阳光下，让我们超越自我，以梦想为指南奋力前行！

学习内容

本章主要讲解使用 Vue CLI 搭建项目，包含的主要内容有 Vue CLI 的简介及安装，Vue CLI 初始化项目的介绍，使用 Vue CLI 搭建品茶轩项目。本章涉及的知识点较多，包括 Node 环境的安装、Vue CLI 的安装、项目的创建、Element UI 的使用、路由的配置、组件的导入、组件之间的通信、项目的打包等。

学习目标

1. 掌握 Vue CLI 的安装方法。
2. 掌握创建项目的方法。
3. 掌握 Element UI 的使用方法。
4. 掌握路由的配置方法。
5. 掌握组件的导入方法。

学习重点

1. 掌握常用的命令行指令。
2. 掌握 Vue CLI 创建项目的目录结构。
3. 掌握 Element UI 的常用组件。
4. 掌握路由的配置方法。

9.1　Vue CLI 简介及安装

9.1.1　Vue CLI 简介

　　Vue CLI 是 Vue 官方出品的快速构建单页应用的脚手架。它的官网地址是 https://cli.vuejs.org/，如图 9-1 所示。Vue CLI 能够帮助开发者自动配置项目结构，集成常用工具，提供热更新等功能，大大简化了 Vue.js 项目的搭建和开发过程。在第 8 章中，我们使用组件和路由完成了茶叶网站的制作，但是从创建过程中可以看到所有的组件都放在一个文件中不利于设计和管理，而使用 Vue CLI 就可以将组件放在每一个单独的 .vue 后缀的文件中。这样就提升了 Web 项目搭建和开发的效率。

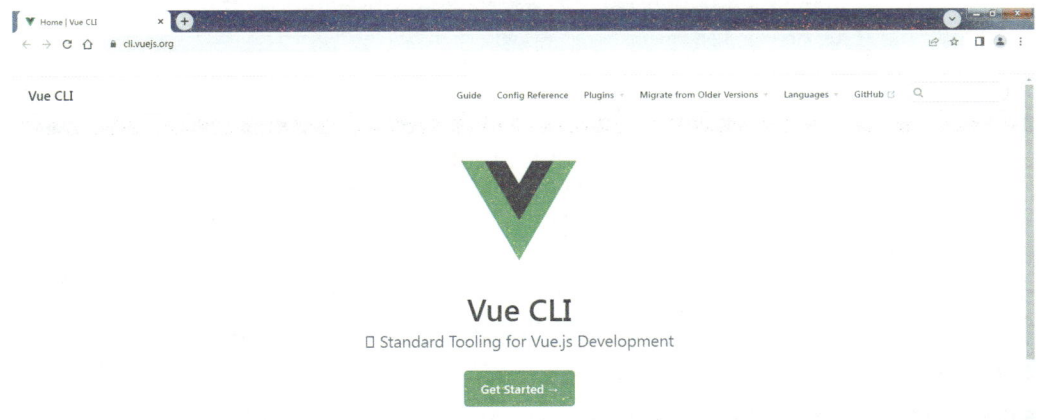

图 9-1　Vue CLI 官网

　　Vue CLI 是一个全局安装的 NPM（node package manager，node 包管理器）包，提供了终端里的 Vue 命令。在安装 Vue CLI 之前必须先安装 Node.js，从而使用 Node.js 的 NPM。Node.js 的网址是 https://nodejs.org/en，如图 9-2 所示。历史版本的下载网址是 https://nodejs.org/en/blog/release，如图 9-3 所示。为了能兼容 Windows 7 操作系统，本章使用的 Node 版本为 v12.18.4。

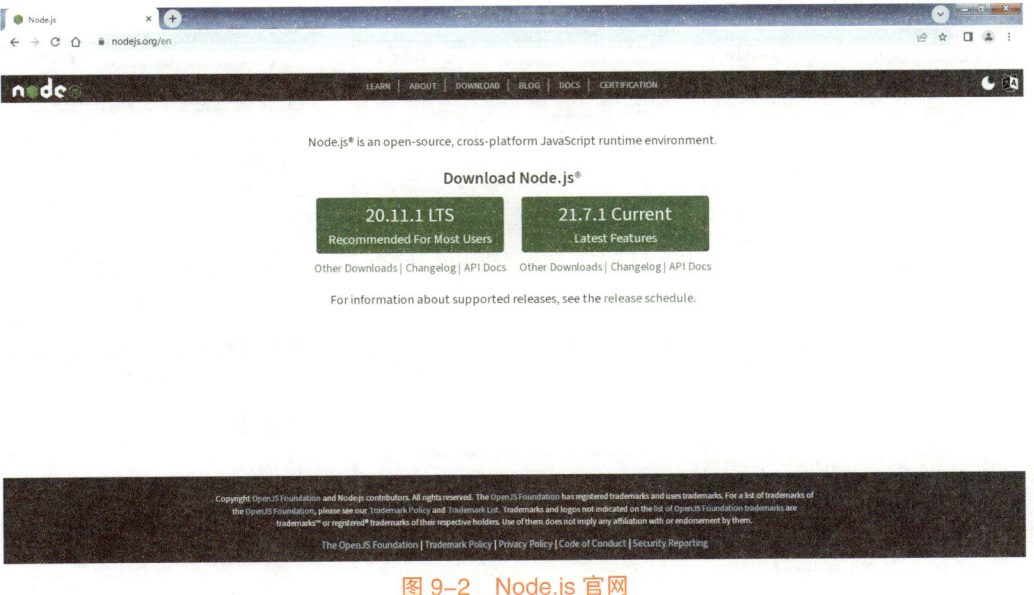

图 9-2　Node.js 官网

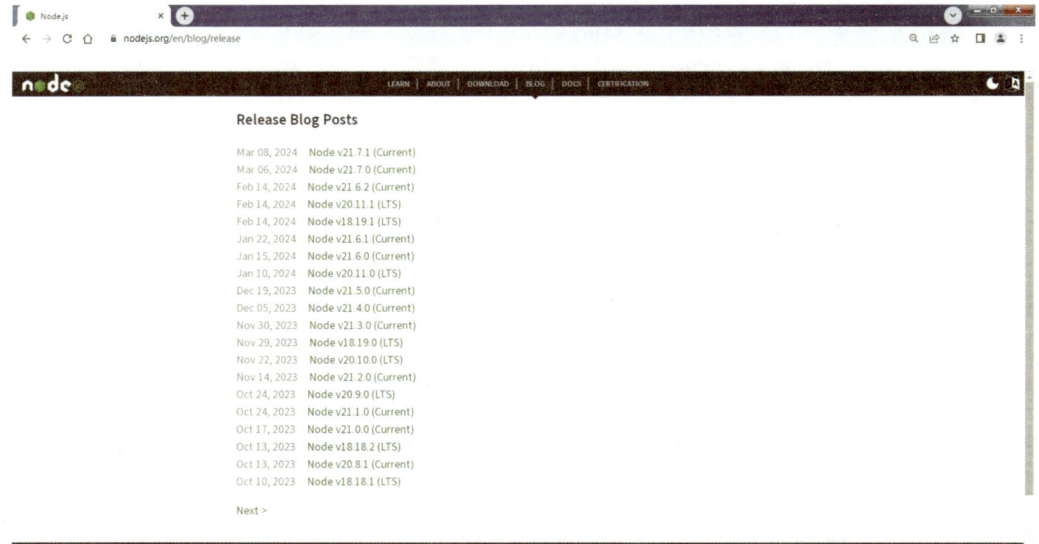

图 9-3　Node 历史版本下载

9.1.2　Node.js 的安装

双击 Node.js 的安装文件，如图 9-4 所示，然后点击 Next 按钮，按提示的步骤进行安装，如图 9-5 ~ 图 9-12 所示。

图 9-4　双击安装

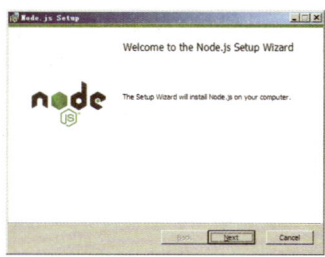

图 9-5　Node.js 安装步骤 1

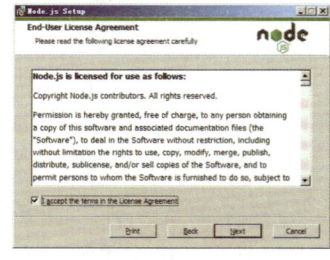

图 9-6　Node.js 安装步骤 2

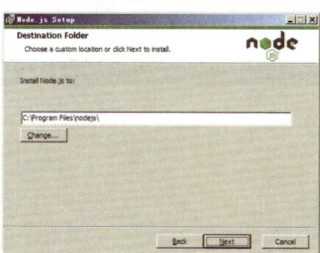

图 9-7　Node.js 安装步骤 3

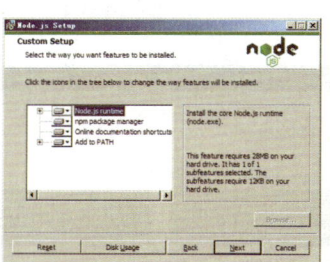

图 9-8　Node.js 安装步骤 4

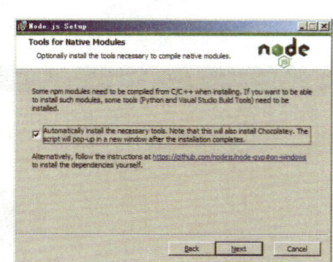

图 9-9　Node.js 安装步骤 5

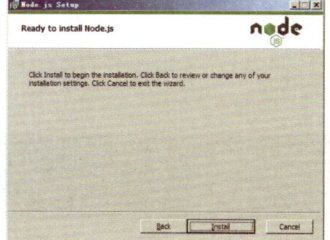

图 9-10　Node.js 安装步骤 6

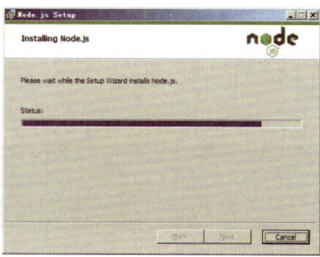

图 9-11　Node.js 安装步骤 7

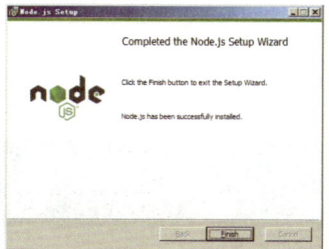

图 9-12　Node.js 安装步骤 8

Node.js 安装完毕以后需要将 Node.js 添加到环境变量中。通过配置环境变量，可以将 Node.js 程序所在的路径添加到系统的搜索路径中，使得用户在任何位置都可以直接执行 Node.js 程序。首先要先获得 Node.js 的安装路径。点击电脑的"开始"按钮，在"所有程序"中找到安装的 Node.js，选中后单击鼠标右键，点击"属性"，如图 9-13 所示。这时会弹出一个对话框，在对话框中的"目标"一栏中将 Node.js 的安装路径复制出来，如图 9-14 所示。接下来在"计算机"上单击鼠标右键，选择"属性"，如图 9-15 所示。然后点击"高级系统设置"，如图 9-16 所示，进入系统设置的对话框。在出现的对话框中再点击右下方的"环境变量（N）…"，如图 9-17 所示，在弹出的对话框中点击"新建"按钮，如图 9-18 所示。然后在新建用户变量对话框中进行设置。变量名处输入 node.js，变量值处将图 9-14 处复制的目标路径复制过来，如图 9-19 所示。

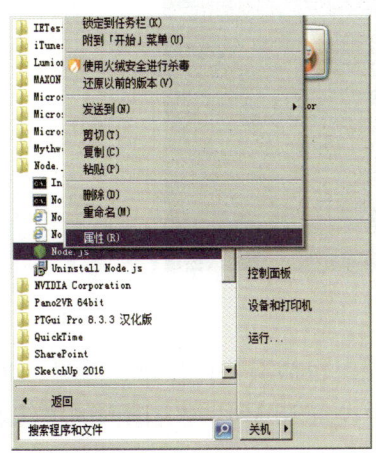
图 9-13　查看 Node.js 属性

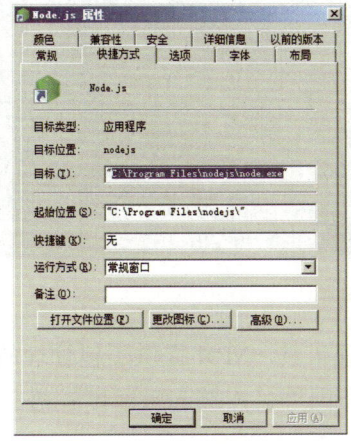
图 9-14　复制目标中的文字

图 9-15　查看"计算机"属性

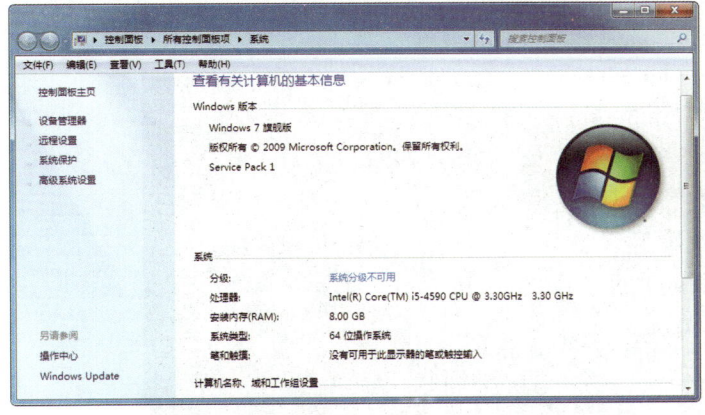
图 9-16　高级系统设置

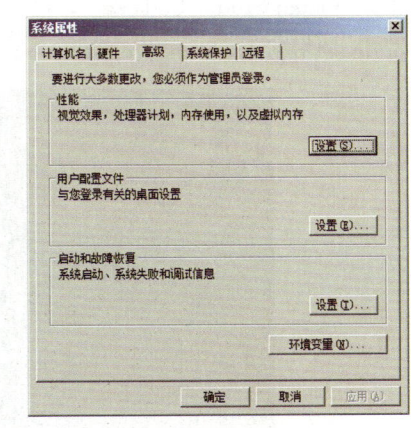
图 9-17　点击环境变量

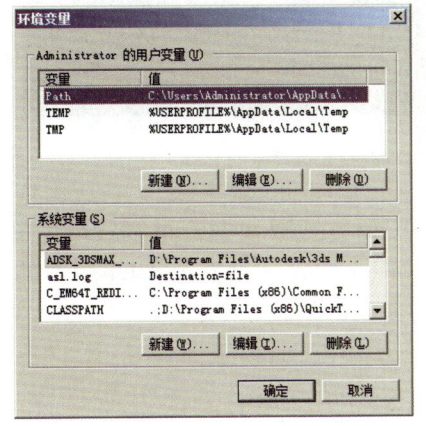
图 9-18　点击新建环境变量

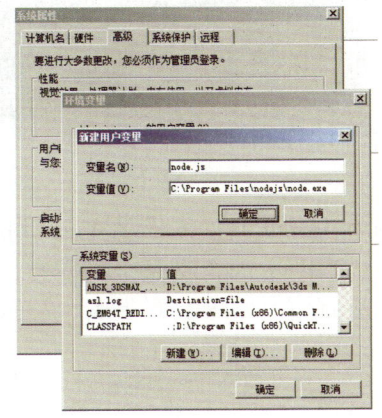
图 9-19　添加环境变量

9.1.3 全局安装 Vue CLI

本章使用的编译器是 Visual Studio Code，简称 VS Code。该编译器的下载网址是 https://code.visualstudio.com/，如图 9-20 所示，点击"Download"按钮可以从网站上下载该编译器，具体的安装这里不再赘述。

全局安装 Vue CLI

图 9-20　Visual Studio Code 的下载网站

NPM 是 Node.js 平台的默认包管理工具。NPM 提供了包管理的功能，包括安装、卸载、更新、查看、搜索和发布等。由于 NPM 的服务器在国外，连接起来速度较慢，下面设置 NPM 代理为国内的镜像服务器，这样管理包的时候速度更快也更稳定。

启动 Visual Studio Code 编译器，点击"终端"菜单下面的"新建终端"。在终端中输入 npm config set registry https://registry.npmmirror.com，如图 9-21 所示。

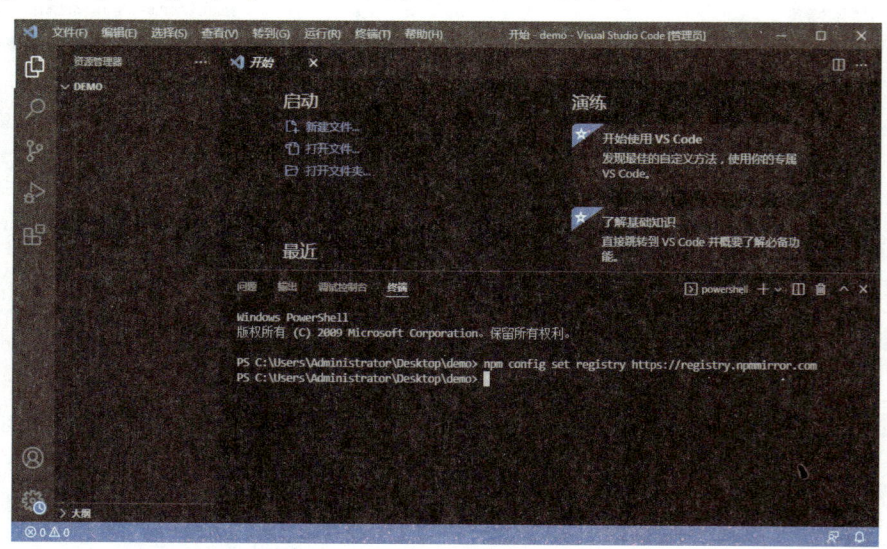

图 9-21　设置 NPM 代理

下面使用 npm install 命令来安装 Vue CLI。在终端中输入 npm install -g '@vue/cli@4.1.1' 进行安装，如图 9-22 所示。这里 -g 的含义是全局安装，@vue/cli@4.1.1 的含义是安装 4.1.1 版本的 Vue CLI，整个安装需要持续一段时间，安装完毕以后的效果如图 9-23 所示。

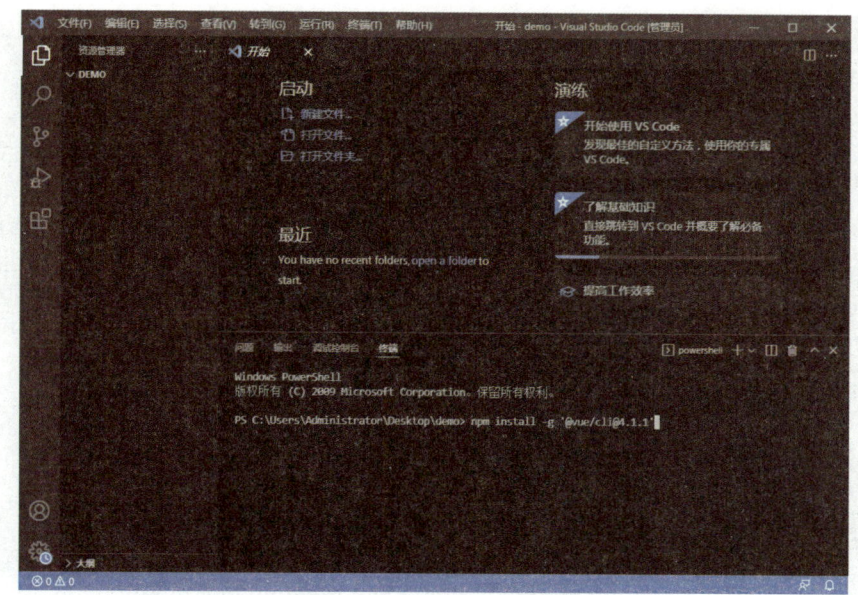

图 9-22　安装 Vue CLI

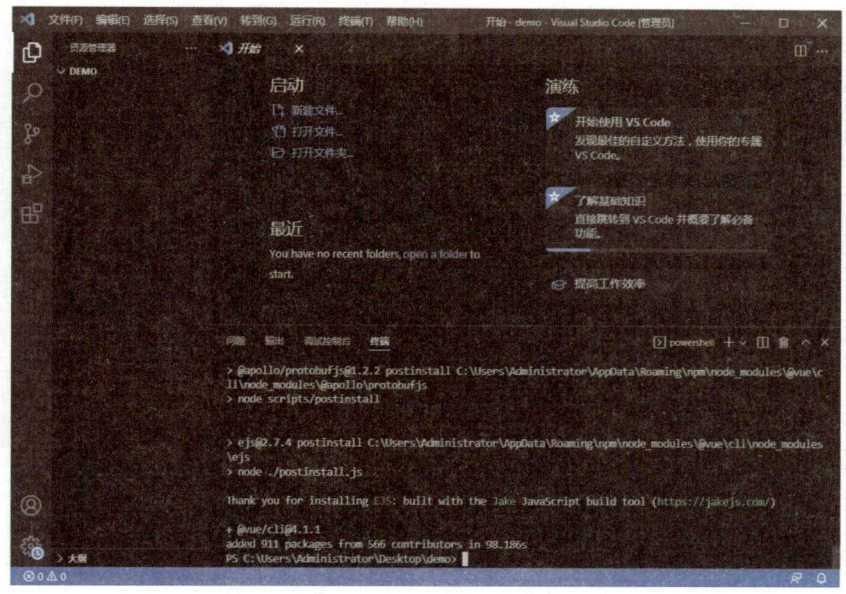

图 9-23　安装完毕

9.1.4　创建项目

在 Visual Studio Code 终端中输入 vue create tea，其中 tea 是项目的名称，点击回车以后出现报错，如图 9-24 所示。这里报错的原因是执行脚本时没有得到 PowerShell 的权限。所以下面要在 PowerShell 中进行权限的设置。

在搜索栏中输入 powershell，以管理员身份打开，如图 9-25 所示，在弹出的窗口中输入 set-ExecutionPolicy RemoteSigned 按回车键，这时会出现询问是否更改，输入 y 然后按回车键，效果如图 9-26 所示。这时候再在 Visual Studio Code 终端中输入 vue create tea 就可以正常创建项目了，如果还是报错可以注销或者重启一下计算机以后再创建项目。

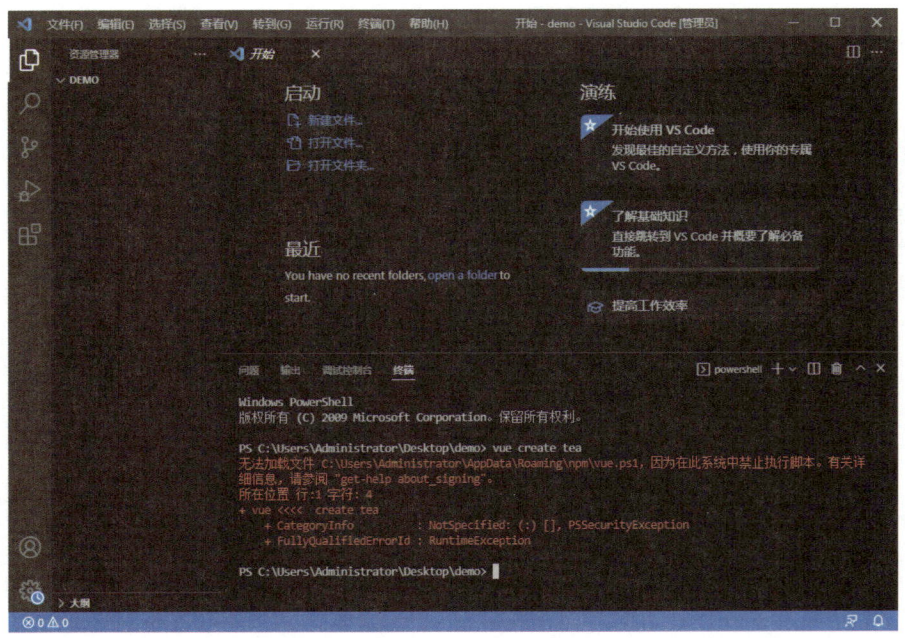

图 9-24　创建项目报错

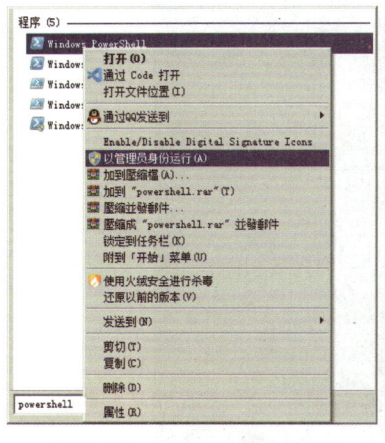

图 9-25　打开 powershell

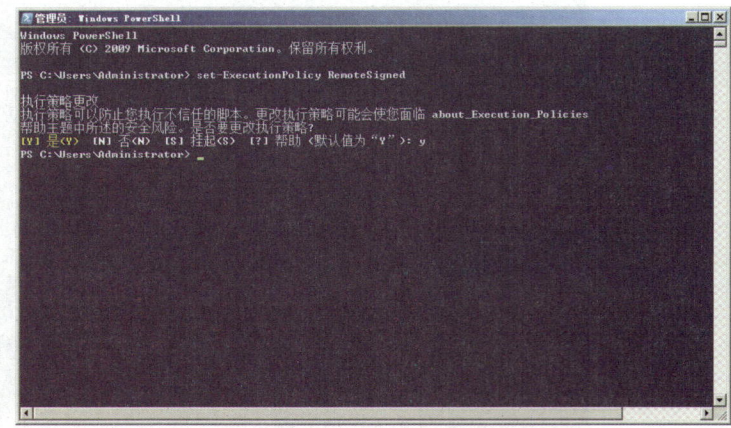

图 9-26　更改 powershell 执行策略

在创建项目时选择 default，进行默认安装，如图 9-27 所示。安装完毕以后出现提示代码，如图 9-28 所示。cd tea 的含义是进入项目目录中，npm run serve 的作用是启动项目。项目启动完毕以后会出现链接的地址如图 9-29 所示，按 Ctrl 键点击链接可以在浏览器中查看项目效果，如图 9-30 所示。

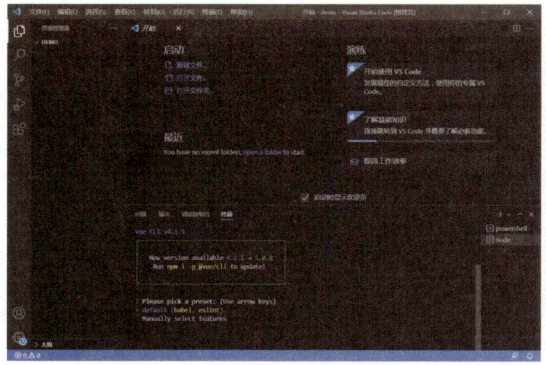

图 9-27　default 默认安装

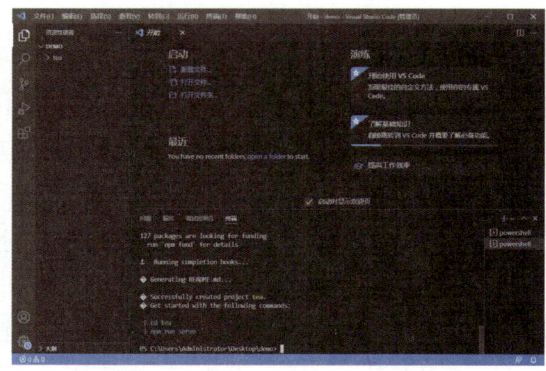

图 9-28　安装完毕以后出现提示代码

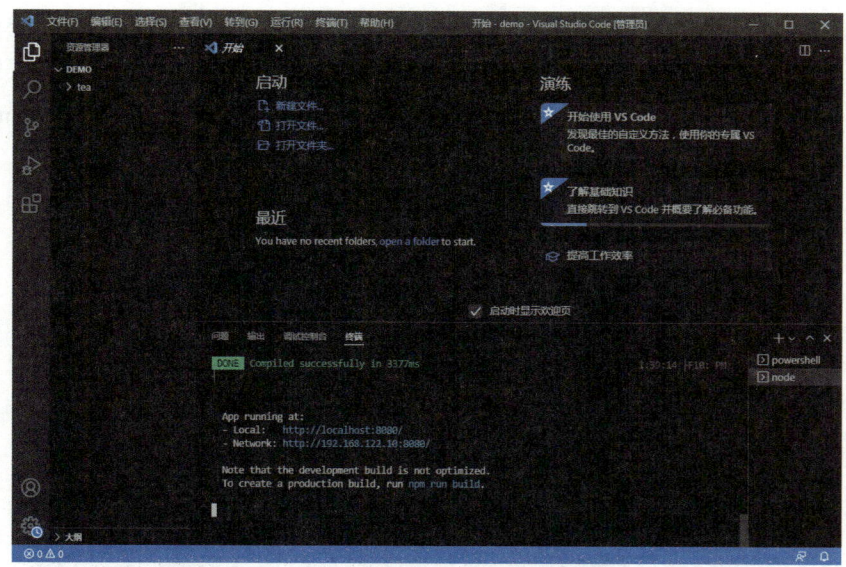

图 9-29　项目链接地址

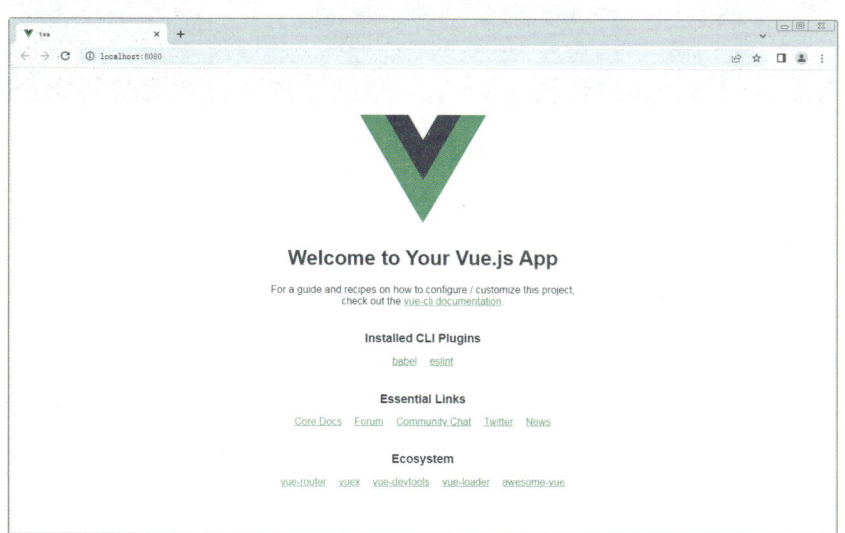

图 9-30　在浏览器中查看项目

9.2　Vue CLI 初始化项目介绍

9.2.1　Vue CLI 目录结构

项目创建以后，在 VS Code 左侧的资源管理器中展开 tea 目录，效果如图 9-31 所示。在初始化的项目中有 node_modules 文件夹、public 文件夹、src 文件夹。

node_modules 是项目依赖目录。node_modules 文件夹中放置的该项目所有安装的包，在传统的 Web 开发中使用 <script></script> 标签进行 JS 文件的引入，但是在 Vue CLI 项目中 JS 文件的引入需要使用 npm install 的方式进行安装，安装以后会在 node_modules 文件夹中出现。

public 是静态资源目录。在 public 文件夹中有 .ico 文件和 index.html 文件。index.html 是 webpack 打包时使用的模板文件。

src 是项目源代码目录。src 文件夹中有 assets 文件夹、components 文件夹、App.vue 和 main.js。assets 文件夹一般放置项目的静态资源，如图片、全局 CSS 文件等。components 文件夹中放置组件文件。在 Vue CLI 中组件是一个独立的以 .vue 为后缀的文件。App.vue 是项目的根组件。main.js 是项目的入口文件，一般在这个文件中引入项目需要的依赖和路由等。

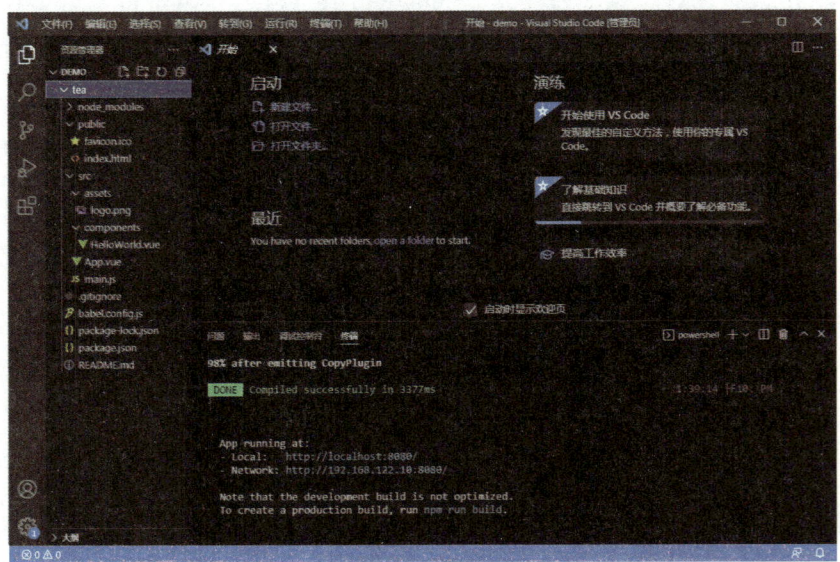

图 9-31　项目目录

9.2.2　Vue CLI 页面结构

　　在 Vue CLI 中的一个 .vue 文件通常包含三个主要部分：模板（template）、脚本（script）和样式（style）。在 JavaScript 中，模块是一种封装了特定功能的代码的单元。通过 export default 关键字来导出模块的默认成员，通过 import 关键字导入模块的成员。下面以 HelloWorld.vue 文件和 App.vue 文件为例进行讲解。

　　HelloWorld.vue 文件代码如下：

```
<template>
  <div class="hello">
    <h1>{{ msg }}</h1>        使用插值表达式渲染 msg
    <p>
      For a guide and recipes on how to configure / customize this project,<br>
      check out the
      <a href="https://cli.vuejs.org" target="_blank" rel="noopener">vue-cli documentation</a>.
    </p>
    <h3>Installed CLI Plugins</h3>
    <ul>
      <li><a href="https://github.com/vuejs/vue-cli/tree/dev/packages/%40vue/cli-plugin-babel" target="_blank" rel="noopener">babel</a></li>
```

```html
        <li><a href="https://github.com/vuejs/vue-cli/tree/dev/packages/%40vue/cli-plugin-eslint" target="_blank" rel="noopener">eslint</a></li>
      </ul>
      <h3>Essential Links</h3>
      <ul>
        <li><a href="https://vuejs.org" target="_blank" rel="noopener">Core Docs</a></li>
        <li><a href="https://forum.vuejs.org" target="_blank" rel="noopener">Forum</a></li>
        <li><a href="https://chat.vuejs.org" target="_blank" rel="noopener">Community Chat</a></li>
        <li><a href="https://twitter.com/vuejs" target="_blank" rel="noopener">Twitter</a></li>
        <li><a href="https://news.vuejs.org" target="_blank" rel="noopener">News</a></li>
      </ul>
      <h3>Ecosystem</h3>
      <ul>
        <li><a href="https://router.vuejs.org" target="_blank" rel="noopener">vue-router</a></li>
        <li><a href="https://vuex.vuejs.org" target="_blank" rel="noopener">vuex</a></li>
        <li><a href="https://github.com/vuejs/vue-devtools#vue-devtools" target="_blank" rel="noopener">vue-devtools</a></li>
        <li><a href="https://vue-loader.vuejs.org" target="_blank" rel="noopener">vue-loader</a></li>
        <li><a href="https://github.com/vuejs/awesome-vue" target="_blank" rel="noopener">awesome-vue</a></li>
      </ul>
    </div>
</template>

<script>
export default {           // 导出模块的默认成员
  name: 'HelloWorld',
  props: {
    msg: String            // 在 props 中定义 msg 的变量类型为 string
  }
}
```

```
</script>

<!-- Add "scoped" attribute to limit CSS to this component only -->
<style scoped>        ◁── 使用 scoped 属性以后,样式仅对当前文档有效
h3 {
  margin: 40px 0 0;
}
ul {
  list-style-type: none;
  padding: 0;
}
li {
  display: inline-block;
  margin: 0 10px;
}
a {
  color: #42b983;
}
</style>
```

App.vue 文件代码如下:

```
<template>
  <div id="app">
    <img alt="Vue logo" src="./assets/logo.png">
    <HelloWorld msg="Welcome to Your Vue.js App"/>   ◁── 应用 HelloWorld 组件,传入 msg 的值
  </div>
</template>

<script>
import HelloWorld from ´./components/HelloWorld.vue´    ◁── 导入 HelloWorld 组件

export default {
  name: ´App´,
  components: {
    HelloWorld        ◁── 声明使用 HelloWorld 组件
  }
}
</script>

<style>
```

```
#app {
  font-family: Avenir, Helvetica, Arial, sans-serif;
  -webkit-font-smoothing: antialiased;
  -moz-osx-font-smoothing: grayscale;
  text-align: center;
  color: #2c3e50;
  margin-top: 60px;
}
</style>
```

在 Vue CLI 项目中，每个组件内都会存在一个 <style></style> 标签，在开发过程中各组件的样式可能会出现选择器相同的情况。在 webpack 编译的时候，所有组件内的 style 都会合并，相同选择器的样式就会进行覆盖。为了避免这个问题，Vue CLI 使用 scoped 属性来作为样式隔离，避免一个组件的样式对其他组件产生影响。

9.3 使用 Vue CLI 完成品茶轩项目

9.3.1 品茶轩项目介绍

在第 8 章中我们使用组件完成了品茶轩网站的制作，而下面我们将使用 Vue CLI 来完成品茶轩项目。该项目中涉及的知识点有项目依赖的安装、路由的配置、组件的建立、组件的通信、less 的使用等。

在第 8 章中我们使用了 Bootstrap 框架，在这一章里我们使用 Element UI 框架。Element UI 是一款为 Vue.js 设计的 UI 库，它的功能强大，提供了数据表格、表单、导航菜单、时间线等多种丰富的组件，方便开发者快速搭建页面。由于两个框架的差异性，所以两个项目的效果相似但不完全相同。如图 9-32 所示是使用 Element UI 框架制作的品茶轩项目。

图 9-32 使用 Element UI 框架制作的品茶轩项目

9.3.2 品茶轩项目制作

项目制作步骤 1：初始化一个干净的 App.vue

在这一步中首先将 components 文件夹下面的 HelloWorld.vue 删除，然后对 App.vue 进行初始化处理，使 App.vue 中只保留最基本的结构。具体代码如下：

```html
<template>
    <div id="app">
    </div>
</template>

<script>
export default {
}
</script>

<style>
</style>
```

品茶轩项目制作

项目制作步骤 2：安装依赖

在这个项目中需要的依赖有 vue-router、less、less-loader、Element UI。其中，vue-router 是路由文件；less 是一种 CSS 预处理器，相比传统的 CSS，less 的功能更加强大，它可以使用变量、函数、嵌套等高级功能来编写 CSS 代码。由于 .vue 后缀的文件不能在浏览器中直接预览，使用 Vue CLI 编写的项目最后要使用 webpack 来进行打包。less-loader 是 webpack 的一个模块加载器，用于在 webpack 中使用 less。可以将 less 文件转换为 CSS 代码并加载到页面中。Element UI 里面的组件很多，这里要进行按需导入，只将项目中需要的组件进行导入，以减少最后生成的打包文件的体积。

进行项目依赖安装的使用首先一定要确保先进入项目的目录下面。由于在项目的时候是在 demo 文件夹下新建的，所有在 VS Code 中新建终端进入的是 demo 的路径，这里一定要输入 cd tea 先进入 tea 目录下，然后再安装依赖，否则的话就会将依赖安装在 demo 目录下。如图 9-33 所示，使用 npm install vue-router@3.2.0 安装 vue-router。如图 9-34 所示，使用 npm install less@3.0.0 安装 less。如图 9-35 所示，使用 npm install less-loader@5.0.0 安装 less-loader。

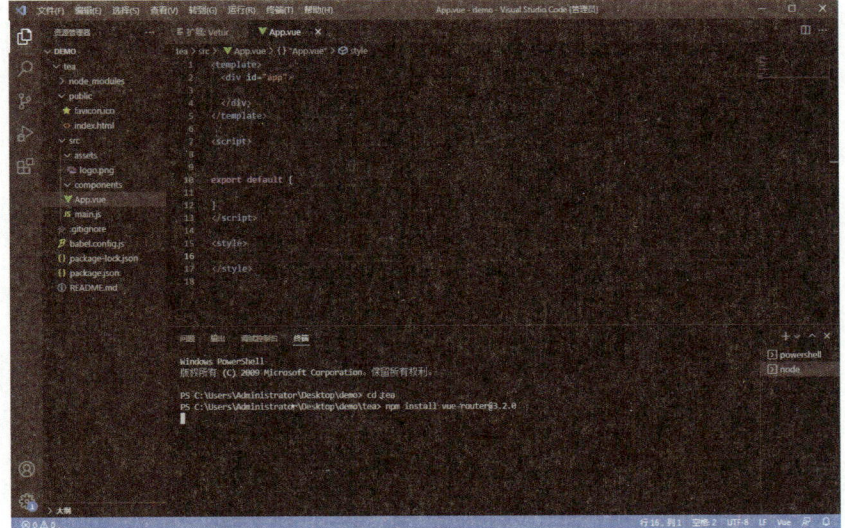

图 9-33　安装 vue-router

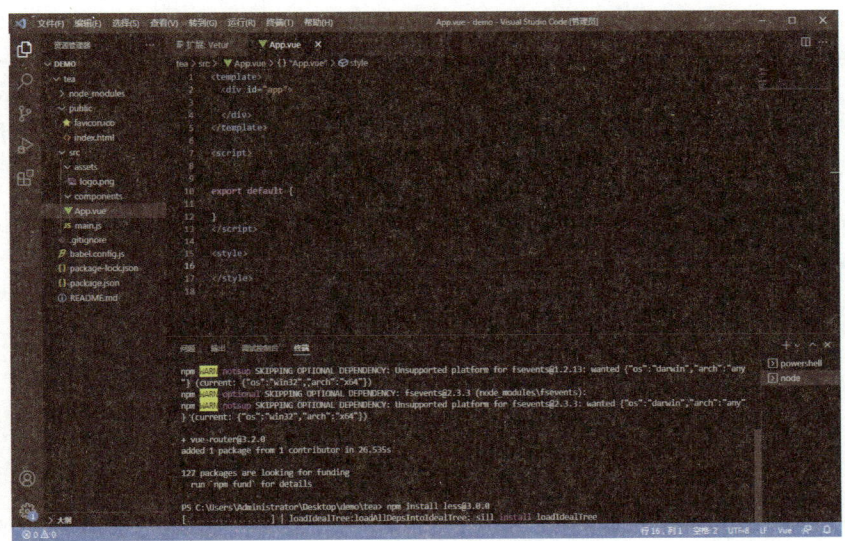

图 9-34　安装 less

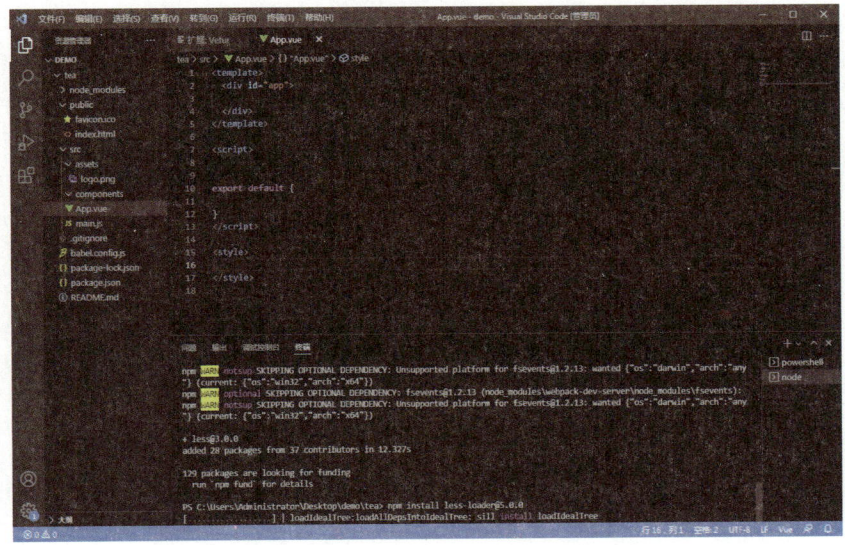

图 9-35　安装 less-loader

Element UI 的安装要复杂一些。Element UI 的网址为 https://element.eleme.cn/#/zh-CN，其网站界面如图 9-36 所示。

图 9-36　Element UI 网站界面

点击"组件"链接进入组件页面，在页面左侧的导航中点击"安装"，页面右侧会有 npm 安装的代码，如图 9-37 所示。点击"快速安装"，页面会有按需导入的设置方法，如图 9-38 所示。在 VS Code 的终端中按照 Element UI 的说明进行安装。首先使用 npm i element-ui-S 安装 Element UI，如图 9-39 所示；然后使用 npm install babel-plugin-component-D 安装 babel-plugin-component，如图 9-40 所示；最后修改项目中的 babel.config.js 文件，按照提示，修改后的 babel.config.js 的代码如下：

```
module.exports = {
  presets: [
    '@vue/cli-plugin-babel/preset'
  ],
  "plugins": [
    [
      "component",
      {
        "libraryName": "element-ui",
        "styleLibraryName": "theme-chalk"
      }
    ]
  ]
}
```

图 9-37　npm 安装 Element UI

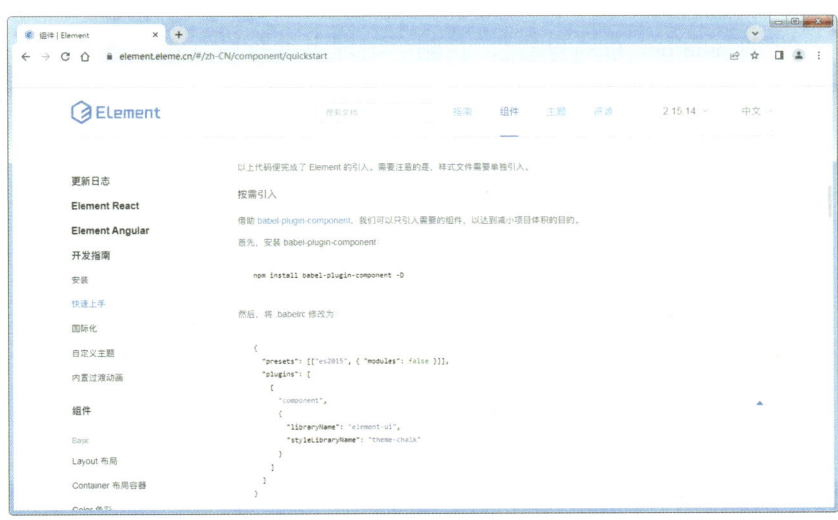

图 9-38　按需导入

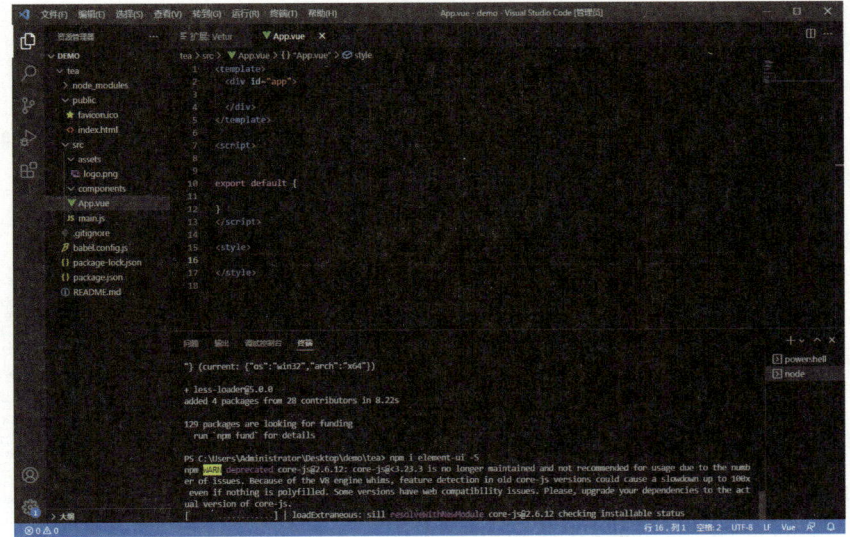

图 9-39　安装 Element UI

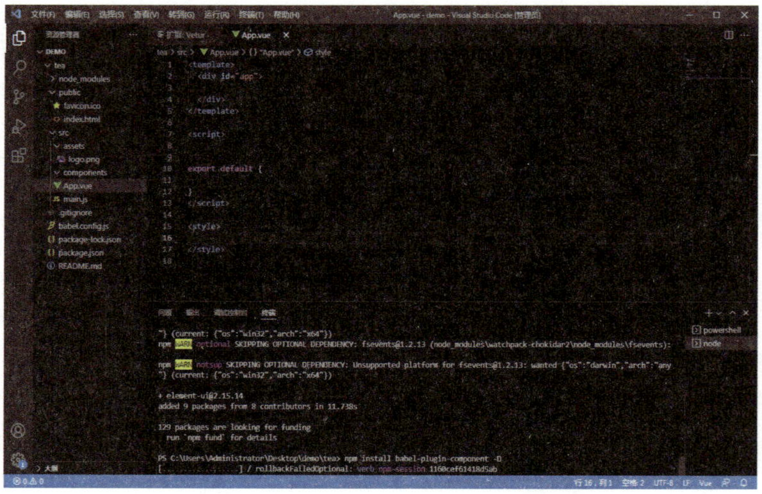

图 9-40 安装 babel-plugin-component

项目制作步骤 3：使用 Element UI 的 Container 布局容器组件

点击左侧导航中的"Container 布局容器"，右侧会出现多种布局方式，当鼠标经过布局效果的下方时会出现"显示代码"。在显示出来的代码中复制第四种布局的代码，将代码粘贴到 App.vue 中。此时，App.vue 的代码如下：

```
<template>
    <div id="app">
        <el-container>
            <el-header>Header</el-header>
            <el-container>
                <el-aside width="200px">Aside</el-aside>
                <el-main>Main</el-main>
            </el-container>
        </el-container>
    </div>
</template>

<script>
    export default {
    }
</script>

<style>
</style>
```

Element UI 的组件都是以 el- 开始的，如 el-container、el-header 等。下面需要在 main.js 中使用 import 按需引入 App.vue 中出现的所有 Element UI 的组件，然后使用 use 方法进行使用。在下面的开发过程中使用到的其他 Element UI 组件都需要在 main.js 中进行导入和使用。此时，main.js 的代码如下：

```
import Vue from 'vue'
```

```
import App from './App.vue'
Vue.config.productionTip = false

import {Container,Header,Aside,Main} from 'element-ui'
Vue.use(Container)
Vue.use(Header)
Vue.use(Aside)
Vue.use(Main)

new Vue({
  render: h => h(App),
}).$mount('#app')
```

在终端中使用 npm run serve 启动项目，效果如图 9-41 所示。

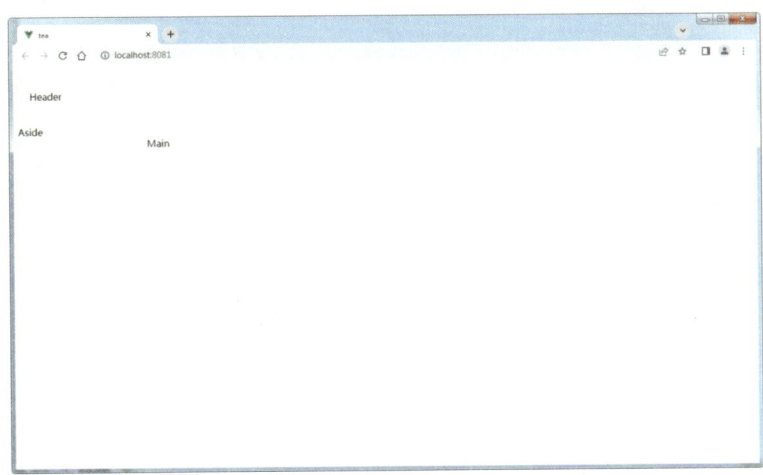

图 9-41 项目页面结构

项目制作步骤 4：新建 Header 组件，在 App.vue 中进行导入

在 VS Code 中的 components 文件夹上单击鼠标右键，新建文件，输入文件名称 Header.vue。在 Header.vue 中进行头部组件的编辑，并使用 less 进行 CSS 的编辑。以下是 Header.vue 的代码。整个项目所用的图片素材放置在 assets 文件夹下面的 images 文件夹中，Header.vue 文件在 components 中，在引入图片的时候要注意路径的正确写法。在 less 中子选择器的 CSS 代码写在父选择器的大括号内部，这种嵌套的写法是 less 的语法才支持的，所以在 style 标签内部要添加 lang="less" 来表示使用 less 语言编写 CSS 代码。

```
<template>
  <div class="header">

  <div class="header_in">
      <img src="../assets/images/logo.png" class="logo" />
      <span class="h4">品茶轩</span>
    </div>
```

注意正确输入图片路径

```
      </div>
    </template>

    <style lang="less">    使用 less 语法
    .header {
      background-color: #d9edf7;
      height: 60px;
      .header_in {    使用嵌套的方式在父选择器内部定义子选择器
        width: 1200px;
        margin: 0 auto;
        height: 60px;
        img {
          float: left;
        }
        span {
          float: left;
          line-height: 60px;
          margin-left: 15px;
        }
      }
    }
    </style>
```

下面在 App.vue 中导入 Header.vue。导入的方法与之前介绍的 HelloWorld.vue 的导入方法相同，首先要使用 import 进行导入，然后在 components 中注册该组件。Element UI 中的组件样式作为一个类使用，如 el-header 组件在类中使用 .el-header{} 进行样式的定义。此时，App.vue 的代码为如下：

```
<template>
  <div id="app">
    <el-container>
      <el-header>
        <Header></Header>    应用 Header 组件
      </el-header>
      <el-container>
        <el-aside width="200px">Aside</el-aside>
        <el-main>Main</el-main>
      </el-container>
    </el-container>
  </div>
</template>
```

```
<script>
import Header from "./components/Header.vue";   // 导入 Header 组件
export default {
  components: {
    Header,   // 注册组件
  }
};
</script>

<style lang="less">
html,body,#app {
  margin: 0;
  padding: 0;
}
.el-header {   // 对 el-header 组件进行样式的设定
  margin: 0;
  padding: 0;
  margin-bottom: 15px;
}
</style>
```

项目的效果如图 9-42 所示。

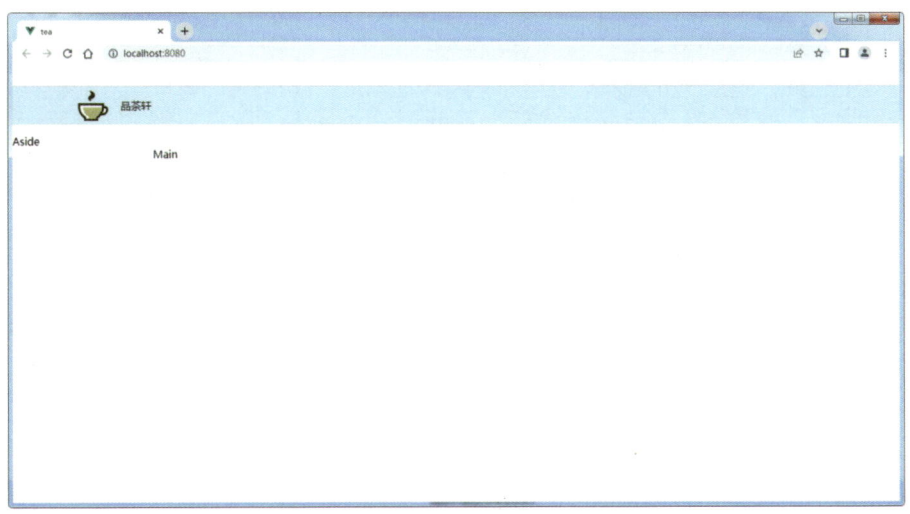

图 9-42　完成 Header 部分

项目制作步骤 5：完成 Aside 部分

在 Aside 部分中要使用 Element UI 的 el-button 组件。点击 Element UI 网站左侧导航中的"Button 按钮"，出现如图 9-43 所示的 el-button 组件。

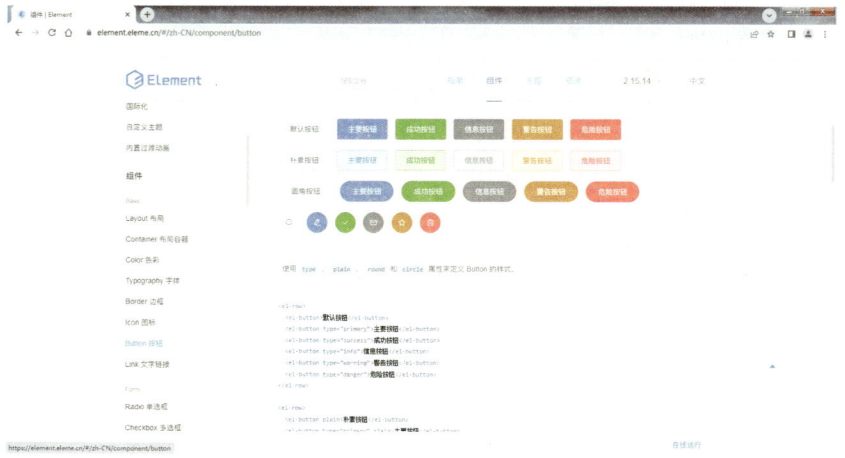

图 9-43　el-button 组件

这里使用"主要按钮"的效果,对应的代码为 <el-button type="primary"> 主要按钮 </el-button>。在使用 el-button 组件的时候一定要确保在 main.js 中导入和使用 Button 组件。

此时,App.vue 文件的代码如下:

```
<template>
  <div id="app">
    <el-container>
      <el-header>
        <Header></Header>
      </el-header>

      <el-container class="cont">
        <el-aside width="200px">
          <ul class="nav">          使用 el-button 按钮
            <li><el-button type="primary">首页</el-button></li>
            <li><el-button type="primary">茶叶分类</el-button></li>
            <li><el-button type="primary">关于</el-button></li>
          </ul>
        </el-aside>
        <el-main>Main</el-main>
      </el-container>
    </el-container>
  </div>
</template>

<script>
import Header from "./components/Header.vue";
export default {
  components: {
```

```
    Header
  }
};
</script>

<style lang="less">
html,
body,
#app {
  margin: 0;
  padding: 0;
}
.el-header {
  margin: 0;
  padding: 0;
  margin-bottom: 15px;
}
.cont {          // 设置 .cont 在页面居中
  width: 1200px;
  height: 600px;
  margin: 0 auto;
}
.el-aside {      // 设置 el-aside 组件的样式
  background-color: #fcf8e3;
  .nav {
    padding: 20px 20px;
    li {
      list-style: none;
      margin-bottom: 20px;
      .el-button {
        width: 160px;
      }
    }
  }
}
.el-main {
  background-color: #dff0d8;
}
</style>
```

此时的页面效果如图9-44所示。

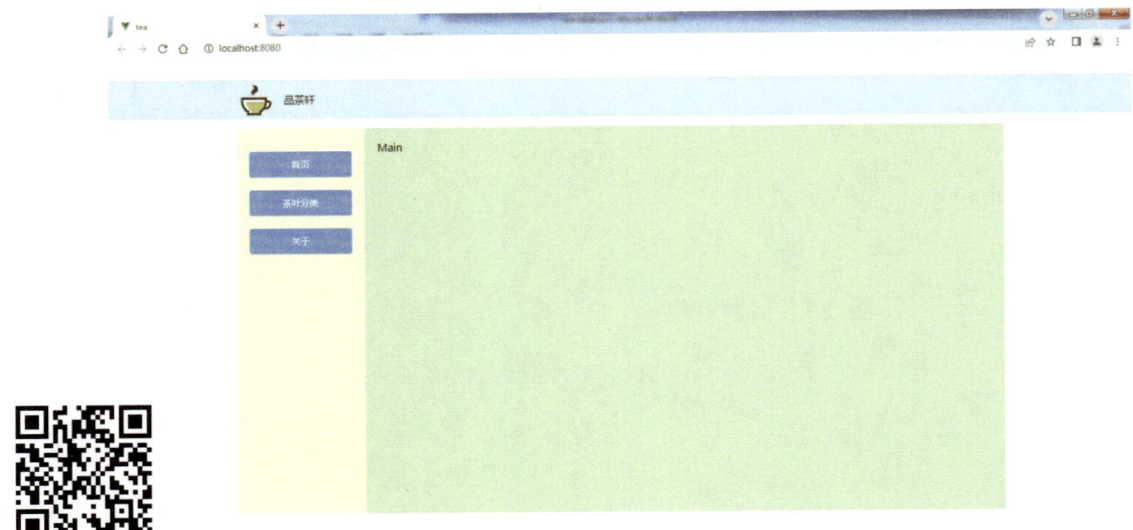

项目制作
步骤6：添加路由

图9-44　完成Aside效果

项目制作步骤6：添加路由

页面右侧的Main部分是要根据左侧点击的链接动态地显示不同的组件，所用Main部分应当使用路由来完成该效果。在步骤2中我们已经将vue-router进行了安装。下面首先在项目的components文件夹上单击鼠标右键，依次新建Home.vue、Tealist.vue、About.vue三个组件，每个组件内部先使用简单的文字进行标记。三个页面的代码依次如下：

```
<template>
    <div>
        Home
    </div>
</template>


<template>
    <div>
        Tealist
    </div>
</template>


<template>
    <div>
        About
    </div>
</template>
```

然后在项目的src文件夹上单击鼠标右键，新建文件，命名为router.js。在router.js中进行路由的引入以及路由的配置。router.js 的代码如下：

```js
import Vue from 'vue'
import Router from 'vue-router'          // 导入三个组件
import home from './components/Home'
import tealist from './components/Tealist'   // 导入三个组件
import about from './components/About'


Vue.use(Router)

const router=new Router({
    routes:[
        {path:'/', redirect:'/home'},
        {path:'/home', component:home},      // 设置路由
        {path:'/tealist', component:tealist},
        {path:'/about', component:about},

    ]
})

export default router
```

下面在 main.js 中导入 vue-router，然后进行设置。main.js 的代码如下：

```js
import Vue from 'vue'
import App from './App.vue'
import router from './router'           // 导入 router.js

Vue.config.productionTip = false
import {Container,Header,Aside,Main,Button} from 'element-ui'
Vue.use(Container)
Vue.use(Header)
Vue.use(Aside)
Vue.use(Main)
Vue.use(Button)

new Vue({
  router,                    // 注册路由
  render: h => h(App),
}).$mount('#app')
```

设置好路由以后修改 App.vue 文件，给按钮添加上 router-link，设置好跳转的路径，同时通过 exact-active-class 属性来设置路由匹配上的样式。另外还需要在 <el-main></el-main> 内部添加上 <router-view></router-view>。此时 App.vue 的代码如下：

```
<template>
  <div id="app">
    <el-container>
      <el-header>
        <Header></Header>
      </el-header>
      <el-container class="cont">
        <el-aside width="200px">
          <ul class="nav">
            <li>
              <el-button type="primary">
                <router-link to="/home" exact-active-class="active" class="link">首页
                </router-link>
              </el-button>
            </li>
            <li>
              <el-button type="primary">
                <router-link to="/tealist" exact-active-class="active" class="link">茶叶分类</router-link>
              </el-button>
            </li>
            <li>
              <el-button type="primary">
                <router-link to="/about" exact-active-class="active" class="link">关于</router-link>
              </el-button>
            </li>
          </ul>
        </el-aside>
        <el-main><router-view></router-view></el-main>
      </el-container>
    </el-container>
  </div>
</template>

<script>
import Header from "./components/Header.vue";
```

```
export default {
  components: {
    Header
  }
};
</script>

<style lang="less">
html,
body,
#app {
  margin: 0;
  padding: 0;
}
.el-header {
  margin: 0;
  padding: 0;
  margin-bottom: 15px;
}
.cont {
  width: 1200px;
  height: 600px;
  margin: 0 auto;
}
.el-aside {
  background-color: #fcf8e3;
  .nav {
    padding: 20px 20px;
    li {
      list-style: none;
      margin-bottom: 20px;
      .el-button {
        width: 160px;
      }
      .link {
        color: white;
        text-decoration: none;
        display: block;
      }
```

```
    .active {
      font-weight: bold;      ← 设置路由匹配时的样式
      color: #e6d17d;
    }
   }
  }
}
.el-main {
  background-color: #dff0d8;
}
</style>
```

添加路由以后的效果如图 9-45 所示。点击"关于"按钮以后的效果如图 9-46 所示。

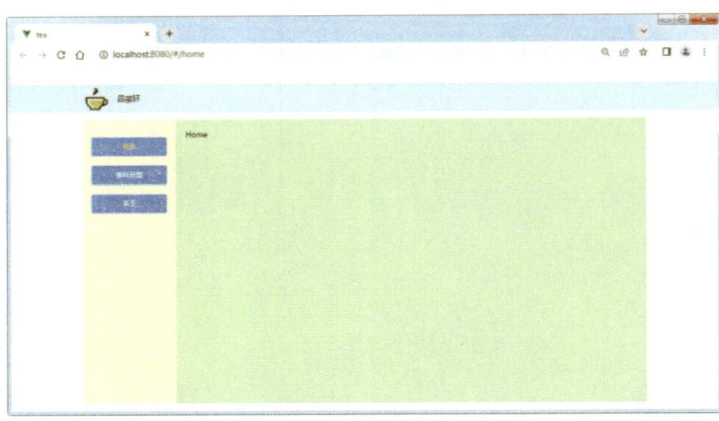

图 9-45　添加路由

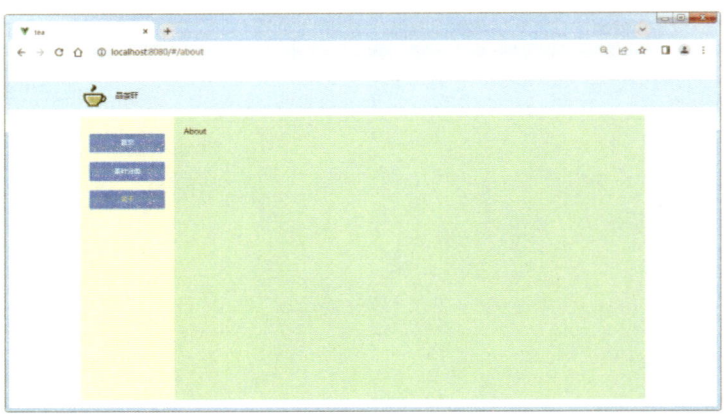

图 9-46　点击"关于"按钮

项目制作步骤 7：完善 Home 等三个组件的内容

在步骤 6 中只是使用文字简单标识了 Home.vue、Tealist.vue、About.vue 三个组件。下面要完善这三个组件。

Home.vue 的代码如下：

```
<template>
  <div class="home">
    <img src="../assets/images/banner.jpg" width="950"/>
```

```
      <p>茶叶，俗称茶，一般包括茶树的叶子和芽。茶叶成分有儿茶素、茶多酚、茶氨酸等。适量喝茶有益健康。
茶叶制成的茶饮料，是世界三大饮料之一。</p>
      <p>茶叶源于中国，茶叶最早是被作为祭品使用的。但从春秋后期就被人们作为菜食，在西汉中期发展为药用，
西汉后期才发展为宫廷高级饮料，普及民间作为普通饮料那是西晋以后的事。发现最早人工种植茶叶的遗迹在浙江余姚
的田螺山遗址，已有6000多年的历史。
      </p>
   </div>
</template>
```

在Tealist.vue中要使用Element UI的Layout布局，点击Element UI左侧导航的"Layout布局"链接，找到右侧的"分栏间隔"，如图9-47所示。Element UI的Layout布局与Bootstrap的栅格的用法相似，不同的是Bootstrap的栅格是12分栏，而Element UI的Layout布局是24分栏，并且可以使用gutter属性来指定每一栏之间的间隔。Layout布局中要使用的组件有el-row、el-col，要将Row和Col导入main.js中并且使用。

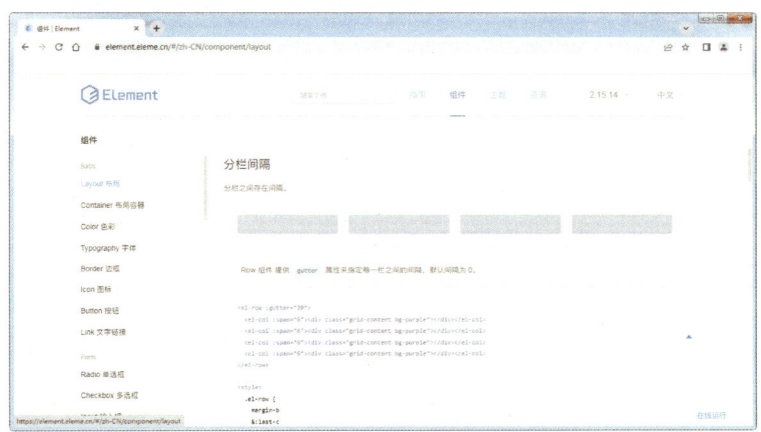

图9-47　Element UI中的Layout布局

Tealist.vue的代码如下：

```
<template>
   <div class="tealist">

<el-row :gutter="20">
      <el-col :span="8">        ← 每一种茶叶占8个分栏宽
          <div class="thumbnail">
             <img src="../assets/images/1.jpg" style="width:100%"/>
             <div class="caption">
                <h3>西湖龙井</h3>
                <p>茶扁平光滑挺直，色泽嫩绿光润，香气鲜嫩清高，滋味鲜爽甘醇</p>
                <p>
                   <el-button type="warning">查看详情</el-button>
                </p>
             </div>
          </div>
```

```
                </el-col>
                <el-col :span="8">
                    <div class="thumbnail">
                        <img src="../assets/images/2.jpg" style="width:100%"/>
                        <div class="caption">
                            <h3>碧螺春</h3>
                            <p>炒成后的干茶条索紧结，翠碧诱人，卷曲成螺，产于春季，故名"碧螺春"</p>
                            <p>
                                <el-button type="warning">查看详情</el-button>
                            </p>
                        </div>
                    </div>
                </el-col>
                <el-col :span="8">
                    <div class="thumbnail">
                        <img src="../assets/images/3.jpg" style="width:100%"/>
                        <div class="caption">
                            <h3>祁门红茶</h3>
                            <p>祁红特绝群芳最，清誉高香不二门，高香美誉，香名远播，美称"红茶皇后"</p>
                            <p>
                                <el-button type="warning">查看详情</el-button>
                            </p>
                        </div>
                    </div>
                </el-col>
            </el-row>

    </div>
</template>

<style scoped>
.thumbnail{ background-color: white; padding: 3px;}
.caption{ padding: 10px;}
</style>
```

在 About.vue 中要使用 Element UI 中的 el-input 组件。要在 main.js 中导入 Input 并且使用。About.vue 的代码如下：

```
<template>
    <div class="about">
        <p>普及茶文化知识是我们的责任和使命，因为茶文化是我们民族独有的文化瑰宝，也是我们国家的重要软实力。通过普及茶文化知识，可以增强人们对传统文化的认同感和自豪感，促进文化传承和创新。茶文化也是中外交流的
```

重要桥梁，通过茶文化的交流可以增进不同国家和地区的友谊和了解。</p>
```
    <h3>联系我们</h3>
    <form>
      <div class="formbox">
        <label for="name">姓名</label>
        <el-input type="text"  id="name" placeholder="请输入姓名" />     ← 使用 el-input
      </div>
      <div class="formbox">
        <label for="idea">建议</label>
        <el-input type="text"  id="idea" placeholder="请输入建议" />
      </div>
      <el-button type="primary" class="btn btn-default">提交</el-button>
    </form>
  </div>
</template>

<style lang="less" scoped>
.formbox{ width: 70%;
.el-input{ margin: 10px 0;}
}

</style>
```

Home.vue 的效果如图 9-48 所示。Tealist.vue 的效果如图 9-49 所示。About.vue 的效果如图 9-50 所示。

图 9-48　Home.vue 效果

图 9-49　Tealist.vue 效果

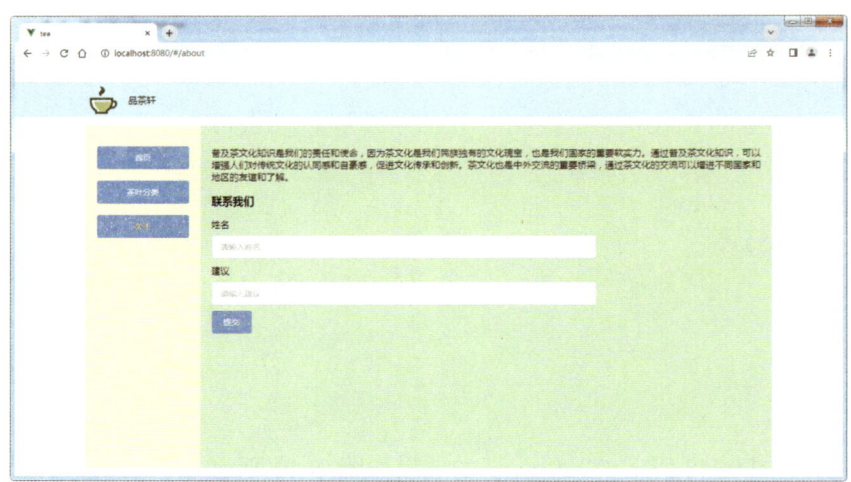

图 9-50　About.vue 效果

项目制作步骤 8：在 App.vue 中添加数据

这一步要在 App.vue 中添加项目所需要的数据，包括导航部分的数据和 Tealist 部分使用的数据。这里要注意在 data 中使用图片路径要使用 require 方法。require 是 Node.js 中的一个方法，它的作用是引入模块、JSON 或本地文件。Vue CLI 默认情况下使用 webpack 进行打包，使用 require 方法可以返回图片编译以后的路径。为了能够实现组件中数据的通信，Tealist.vue 要使用 props 定义 tealist，同时 App.vue 要使用 :tealist="tealist" 来传递数据。App.vue 的代码如下：

```
<template>
  <div id="app">
    <el-container>
      <el-header>
        <Header></Header>
      </el-header>
      <el-container class="cont">

        <el-aside width="200px">
```

```
            <ul class="nav">
                <li v-for="item in nav" :key="item.label">    ← 使用 v-for 循环出导航部分
                    <el-button type="primary">
                            <router-link :to="item.path" exact-active-class="active" class="link">{{item.label}}</router-link>
                    </el-button>
                </li>
            </ul>
        </el-aside>                                              ← 与 Tealist.vue 传递数据
        <el-main><router-view :tealist="tealist"></router-view></el-main>
      </el-container>
    </el-container>
  </div>
</template>

<script>
import Header from "./components/Header.vue";
export default {
  components: {
    Header,
  },
  data() {
    return {
      nav: [    ← 导航部分的数据
        {
          label: "首页",
          path: "/home"
        },
        {
          label: "茶叶分类",
          path: "/tealist"
        },
        {
          label: "关于",
          path: "/about"
        }
      ],
      tealist: [    ← tealist 部分的数据
        {
          id: 0,
```

```
            name: "西湖龙井",
            type: "绿茶",
            place: "浙江省杭州市西湖",
            feature: "茶扁平光滑挺直，色泽嫩绿光润，香气鲜嫩清高，滋味鲜爽甘醇",
            detail:
                "特级西湖龙井，叶底细嫩呈朵。"院外风荷西子笑，明前龙井女儿红。"西湖龙井茶与西湖一样，是人、自然、文化三者的完美结晶，是西湖地域文化的重要载体。",
            smallImg: require('./assets/images/1.jpg'),
            bigImg: require('./assets/images/big1.jpg')   ← 使用require方法定义图片的路径
        },
        {
            id: 1,
            name: "碧螺春",
            type: "绿茶",
            place: "江苏省苏州市太湖洞庭山",
            feature:
                "炒成后的干茶条索紧结，翠碧诱人，卷曲成螺，产于春季，故名"碧螺春"",
            detail:
                "唐朝时就被列为贡品，古人们又称碧螺春为"工夫茶"。此茶冲泡后杯中白云翻滚，清香袭人，是中国的名茶。主要工序为杀青、揉捻、搓团显毫、炒青。",
            smallImg: require('./assets/images/2.jpg'),
            bigImg: require('./assets/images/big2.jpg'),
        },
        {
            id: 2,
            name: "祁门红茶",
            type: "红茶",
            place: "安徽省祁门一带",
            feature:
                "祁红特绝群芳最，清誉高香不二门，高香美誉，香名远播，美称"红茶皇后"",
            detail:
                "茶叶原料选用当地的中叶、中生种茶树"槠叶种"（又名祁门种）制作，是中国历史名茶，著名红茶精品。祁门红茶是红茶中的极品，高香美誉，香名远播，享有盛誉。",
            smallImg: require('./assets/images/3.jpg'),
            bigImg: require('./assets/images/big3.jpg'),
        }
        ]
    };
    }
};
```

```less
</script>

<style lang="less">
html,
body,
#app {
  margin: 0;
  padding: 0;
}
.el-header {
  margin: 0;
  padding: 0;
  margin-bottom: 15px;
}
.cont {
  width: 1200px;
  height: 600px;
  margin: 0 auto;
}
.el-aside {
  background-color: #fcf8e3;
  .nav {
    padding: 20px 20px;
    li {
      list-style: none;
      margin-bottom: 20px;
      .el-button {
        width: 160px;
      }
      .link {
        color: white;
        text-decoration: none;
        display: block;
      }
      .active {
        font-weight: bold;
        color: #e6d17d;
      }
    }
  }
```

```
    }
  }
  .el-main {
    background-color: #dff0d8;
  }
</style>
```

Tealist.vue 的代码如下:

```
<template>
  <div class="tealist">
    <el-row :gutter="20">
      <el-col :span="8" v-for="item in tealist" :key="item.id">
        <div class="thumbnail">
          <img :src="item.smallImg" style="width:100%" />
          <div class="caption">
            <h3>{{item.name}}</h3>
            <p>{{item.feature}}</p>
            <p>
              <el-button type="warning">查看详情</el-button>
            </p>
          </div>
        </div>
      </el-col>
    </el-row>
  </div>
</template>

<script>
export default {
  props: ["tealist"]
};
</script>

<style scoped>
.thumbnail {
  background-color: white;
  padding: 3px;
}
.caption {
  padding: 10px;
}
```

注释:
- 循环 tealist
- 使用 props 定义 tealist

```
        </style>
```

项目制作步骤 9：添加详情组件 Detail.vue

在项目的 components 文件夹上单击鼠标右键，新建文件，命名为 Detail.vue。在 router.js 文件中添加 detail 的路由，同时在 Tealist.vue 中点击"查看详情"的链接可以实现路由传参，具体的实现方法在第 8 章中已经讲解，这里不再赘述。

detail.vue 的代码如下：

```
<template>
  <div class="detail">
    <img :src="tea.bigImg" class="img-responsive" width="900" height="320" />
    <h3>名称：{{tea.name}}</h3>
    <p>
      <b>类型：</b>
      {{tea.type}}
    </p>
    <p>
      <b>产地：</b>
      {{tea.place}}
    </p>
    <p>
      <b>形态：</b>
      {{tea.feature}}
    </p>
    <p>
      <b>特色：</b>
      {{tea.detail}}
    </p>
  </div>
</template>

<script>
export default {
  data() {
    return {
      tea: "",
      id: ""
    };
  },
  methods: {
    getDetail() {
      var result = this.tealist.filter(item => {
```

```
      return item.id == this.id;
    });
    this.tea = result[0];
   }
  },
  mounted() {
   this.id = this.$route.query.id;
   this.getDetail();
  },
  props: ["tealist"]
};
</script>
```

router.js 的代码如下：

```
import Vue from 'vue'
import Router from 'vue-router'
import home from './components/Home'
import tealist from './components/Tealist'
import about from './components/About'
import detail from './components/Detail'

Vue.use(Router)

const router=new Router({
    routes:[
        {path:'/', redirect:'/home'},
        {path:'/home', component:home},
        {path:'/tealist', component:tealist},
        {path:'/about', component:about},
        {path:'/detail', component:detail},   ← 在路由中添加 /detail


    ]
})

export default router
```

Tealist.vue 的代码如下：

```
<template>
  <div class="tealist">
    <el-row :gutter="20">
      <el-col :span="8" v-for="item in tealist" :key="item.id">
        <div class="thumbnail">
```

```html
          <img :src="item.smallImg" style="width:100%" />
          <div class="caption">
            <h3>{{item.name}}</h3>
            <p>{{item.feature}}</p>
            <p>
              <el-button type="warning" @click="toDetail(item.id)">查看详情</el-button>
            </p>
          </div>
        </div>
      </el-col>
    </el-row>
  </div>
</template>

<script>
export default {
  props: ["tealist"],
  methods: {
    toDetail(teaId) {
      this.$router.push({
        path: "/detail",
        query: {
          id: teaId
        }
      });
    }
  }
};
</script>

<style scoped>
.thumbnail {
  background-color: white;
  padding: 3px;
}
.caption {
  padding: 10px;
}
</style>
```

在 Tealist 中点击"查看详情"按钮就可以跳转到详情页，显示对应的内容，其效果如图 9-51 所示。

图 9-51　跳转详情页

项目制作步骤 10：添加 Footer.vue

在项目的 components 文件上单击鼠标右键，新建文件，命名为 Footer.vue。编辑 Footer.vue，并且导入 App.vue 中。

Footer.vue 的代码如下：

```
<template>
  <div class="footer">
    <p>品茶轩版权所有&copy;2023</p>
  </div>
</template>

<style scoped>
.footer {
  line-height: 60px;
  margin-top: 15px;
  background-color: #d9edf7;
  line-height: 60px;
  text-align: center;
}
</style>
```

App.vue 的代码如下：

```
<template>
  <div id="app">
    <el-container>
      <el-header>
        <Header></Header>
```

```html
      </el-header>
      <el-container class="cont">
        <el-aside width="200px">
          <ul class="nav">
            <li v-for="item in nav" :key="item.label">
              <el-button type="primary">
                <router-link :to="item.path" exact-active-class="active"
                  class="link">{{item.label}}</router-link>
              </el-button>
            </li>
          </ul>
        </el-aside>
        <el-main><router-view :tealist="tealist"></router-view></el-main>
      </el-container>
    </el-container>   <-- 应用 Footer 组件
    <Footer></Footer>
  </div>
</template>
```

```js
<script>
import Header from "./components/Header.vue";
import Footer from "./components/Footer.vue";   <-- 导入 Footer.vue
export default {
  components: {
    Header,
    Footer,   <-- 注册 Footer 组件
  },
  data() {
    return {
      nav: [
        {
          label: "首页",
          path: "/home"
        },
        {
          label: "茶叶分类",
          path: "/tealist"
        },
        {
          label: "关于",
```

```
              path: "/about"
            }
          ],
          tealist: [
            {
              id: 0,
              name: "西湖龙井",
              type: "绿茶",
              place: "浙江省杭州市西湖",
              feature: "茶扁平光滑挺直，色泽嫩绿光润，香气鲜嫩清高，滋味鲜爽甘醇",
              detail:
                "特级西湖龙井，叶底细嫩呈朵。"院外风荷西子笑，明前龙井女儿红。"西湖龙井茶与西湖一样，是人、自然、文化三者的完美结晶，是西湖地域文化的重要载体。",
              smallImg: require('./assets/images/1.jpg'),
              bigImg: require('./assets/images/big1.jpg')
            },
            {
              id: 1,
              name: "碧螺春",
              type: "绿茶",
              place: "江苏省苏州市太湖洞庭山",
              feature:
                "炒成后的干茶条索紧结，翠碧诱人，卷曲成螺，产于春季，故名"碧螺春"",
              detail:
                "唐朝时就被列为贡品，古人们又称碧螺春为"工夫茶"。 此茶冲泡后杯中白云翻滚，清香袭人，是中国的名茶。主要工序为杀青、揉捻、搓团显毫、炒青。",
              smallImg: require('./assets/images/2.jpg'),
              bigImg: require('./assets/images/big2.jpg'),
            },
            {
              id: 2,
              name: "祁门红茶",
              type: "红茶",
              place: "安徽省祁门一带",
              feature:
                "祁红特绝群芳最，清誉高香不二门，高香美誉，香名远播，美称"红茶皇后"",
              detail:
                "茶叶原料选用当地的中叶、中生种茶树"楮叶种"（又名祁门种）制作，是中国历史名茶，著名红茶精品。祁门红茶是红茶中的极品，高香美誉，香名远播，享有盛誉。",
              smallImg: require('./assets/images/3.jpg'),
```

```
            bigImg: require('./assets/images/big3.jpg'),
        }
      ]
    };
  }
};
</script>

<style lang="less">
html,
body,
#app {
  margin: 0;
  padding: 0;
}
.el-header {
  margin: 0;
  padding: 0;
  margin-bottom: 15px;
}

.cont {
  width: 1200px;
  height: 600px;
  margin: 0 auto;
}
.el-aside {
  background-color: #fcf8e3;
  .nav {
    padding: 20px 20px;
    li {
      list-style: none;
      margin-bottom: 20px;
      .el-button {
        width: 160px;
      }
      .link {
        color: white;
        text-decoration: none;
        display: block;
```

```
        }
        .active {
          font-weight: bold;
          color: #e6d17d;
        }
      }
    }
  }
  .el-main {
    background-color: #dff0d8;
  }
}
</style>
```

项目制作步骤 11：打包项目

在 Vue CLI 中，每一个组件都有自己的 template、CSS 和 JavaScript。在大型应用程序中，文件的数量是很多的，如果所有的文件都在浏览器中一次性加载，耗时会很久，也可能会导致浏览器的崩溃。通过打包项目可以解决上述的问题。打包可以合并多种类型的文件，如 JavaScript、CSS 和 HTML 等，从而减少 HTTP 请求的次数，加快应用程序的加载速度。在 Vue CLI 中，最常用的打包工具是 webpack。

在打包之前首先要设置打包的配置文件。在项目的根目录 tea 文件夹上单击鼠标右键，新建文件，命名为 vue.config.js。在 vue.config.js 中进行打包的配置，每一个项目根据不同的要求配置也会不同。下面是基本的配置，具体的代码如下：

```
module.exports = {
    publicPath: './',    // 基本路径
    outputDir: 'dist',   // 构建时的输出目录
    assetsDir: 'static', // 放置静态资源的目录
    indexPath: 'index.html', // html 的输出路径
    filenameHashing: true, // 文件名哈希值
    lintOnSave: false, // 是否在保存的时候使用 `eslint-loader` 进行检查。
    runtimeCompiler: false,
    transpileDependencies: [], // babel-loader 默认会跳过 node_modules 依赖。
    productionSourceMap: false, // 是否为生产环境构建生成 source map
    configureWebpack: () => { }, //调整内部的 webpack 配置
    chainWebpack: () => { },
    // 配置 webpack-dev-server 行为。
    devServer: {
      open: true, // 编译后默认打开浏览器
      host: '0.0.0.0',   // 域名,设置为 0.0.0.0 则所有的地址均能访问
      port: 8080,  // 端口
      https: false,  // 是否 https
      // 显示警告和错误
      overlay: {
```

```
        warnings: false,
        errors: true
    },
  }
}
```

接下来新建终端，使用 cd tea 进入 tea 目录下，输入 npm run build 进行项目打包，如图 9-53 所示。打包完成以后会在项目的根目录下出现一个 dist 文件夹，dist 文件夹内部的结构如图 9-54 所示，这些就是打包以后生成的文件。在图中右侧显示的 js 文件就是将多个文件中的 js 代码合并在一起以后生成的 js 文件。dist 文件夹下面的 index.html 是打包以后的首页文件，双击打开以后的效果如图 9-55 所示。

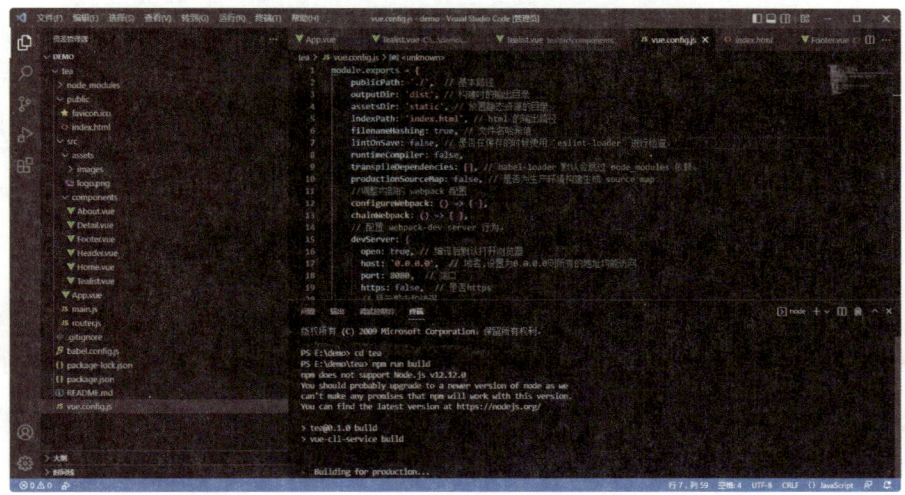

图 9-53　npm run build 打包

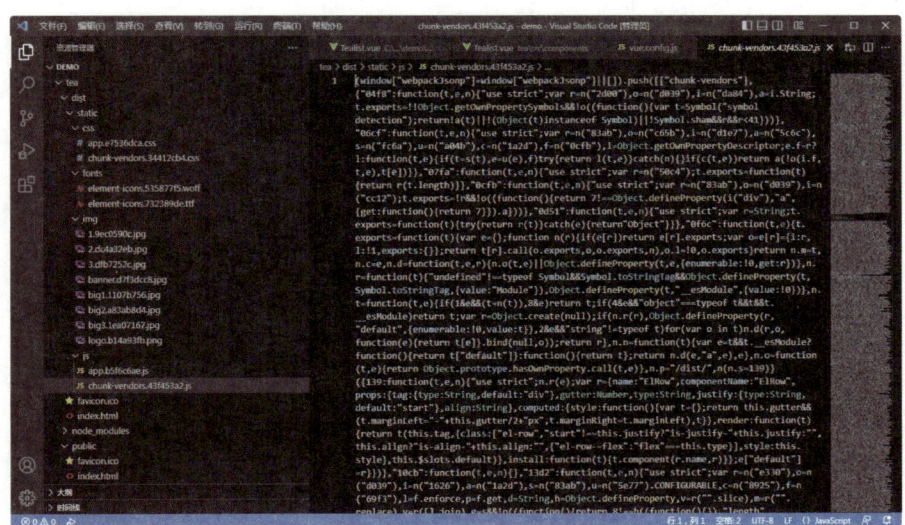

图 9-54　dist 文件夹

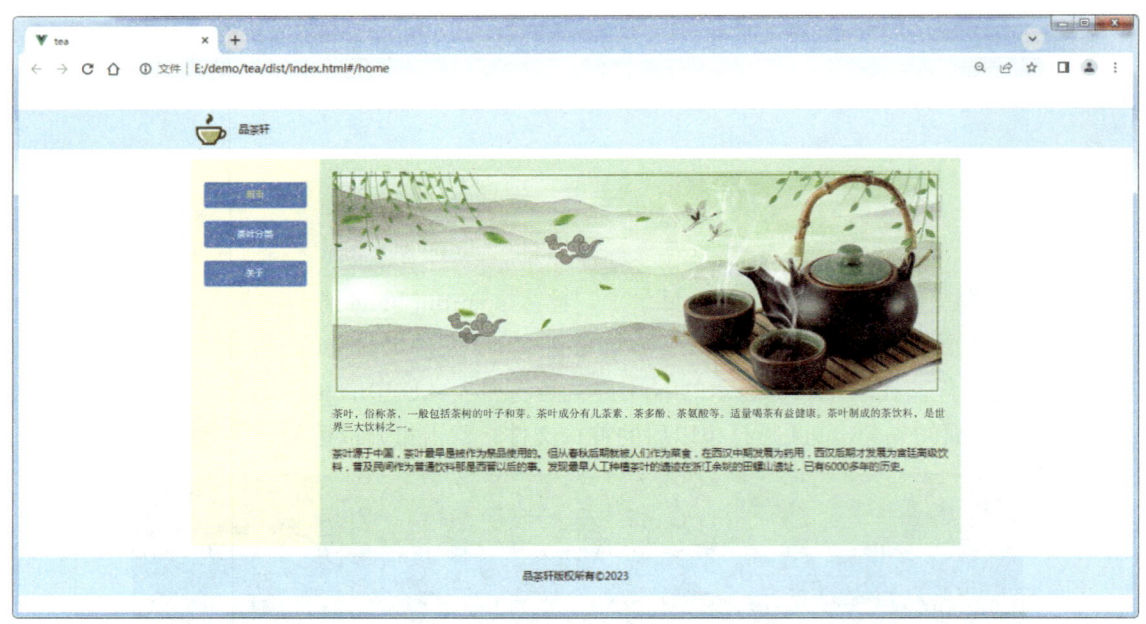

图 9-55　项目打包以后的首页

课后练习题

1. 在 Vue CLI 中组件文件的后缀是_____，一个组件通常由三部分组成，分别是_____、_____、_____。

2. Node.js 平台的默认包管理工具是_____，它提供了包管理的多种功能。

3. 使用 Vue CLI 创建项目使用的命令是_____，启动项目使用的命令是_____。

4. 在 Vue CLI 创建项目中，项目的根组件的是_____，项目入口的 js 文件是_____。

5. Vue CLI 中使用了_____属性，来避免一个组件的样式对其他组件产生影响。

参考文献

[1] 张益珲. 循序渐进 Vue.js 3 前端开发实战［M］. 北京：清华大学出版社，2021.

[2] 张益珲，曹艳琴. 循序渐进 Vue.js 3.x 前端开发实战［M］. 北京：清华大学出版社，2023.

[3] 李小威. Vue.js 3.0 从入门到精通：视频教学版［M］. 北京：清华大学出版社，2021.

[4] 吕云翔，江一帆. Vue 3.0 从入门到实战：微课视频版［M］. 北京：清华大学出版社，2022.

[5] 袁龙. Vue.js 3 企业级项目开发实战：微课视频版［M］. 北京：清华大学出版社，2023.

第 1 章练习题参考答案

1. Model-View-ViewModel，Model，View Model，View
2. <script>
3. el，data，methods
4. createApp，mount
5. 函数返回值

第 2 章练习题参考答案

1. 数字型、字符型、布尔型、数组类型、对象类型
2. 单向绑定，双向绑定，{{ }}
3. this
4. computed
5. watch，computed
6. created
7. mounted

第 3 章练习题参考答案

1. v-text，v-cloak
2. v-on，@
3. prevent，stop，once
4. v-if，v-show，v-show
5. v-for
6. v-bind，v-model

第 4 章练习题参考答案

1. unshift，push，unshift，push
2. shift，pop，shift，pop
3. splice
4. forEach，findIndex，indexOf

第 5 章练习题参考答案

1. transition
2. .v-enter-from，.v-enter-to，.v-enter-active，.v-leave-from，.v-leave-to，.v-leave-active
3. .v-enter 和 .v-leave
4. name
5. transition-group，:key

第 6 章练习题参考答案

1. 请求地址，请求成功的回调函数，请求失败的回调函数
2. 请求地址？参数 = 参数值，params
3. split，replace，slice

第 7 章练习题参考答案

1. template，methods，data
2. 模板字符串，反引号
3. <say-hello></say-hello>
4. :is
5. props
6. ref，自定义事件
7. $emit
8. 默认插槽，具名插槽，作用域插槽
9. name

第 8 章练习题参考答案

1. SPA
2. router-link，router-view，变量
3. #
4. <a>
5. children
6. .router-link-active，.router-link-exact-active。
7. query，params
8. window.sessionStorage

第 9 章练习题参考答案

1. vue，template，CSS，JavaScript
2. npm
3. vue create 项目名称，npm run serve
4. App.vue，main.js
5. scoped